T0192717

Martin Tajmar

Advanced
Space Propulsion Systems

SpringerWienNewYork

Dipl. Ing. Dr. Martin Tajmar
ARC Seibersdorf Research GmBH, Seibersdorf, Austria

*This book was generously supported by the
Austrian Research Centers*

Typesetting: Thomson Press Ltd., Chennai, India
Printing: A. Holzhausen Nfg., A-1140 Wien

Printed on acid-free and chlorine-free bleached paper
SPIN: 10888345

With 121 (partly coloured) Figures

CIP-data applied for

ISBN 3-211-83862-7 Springer-Verlag Wien New York

Contents

Acknowledgement

I'm very grateful to Erich Kny and Wolfgang Renner from the Austrian Research Centers for their support and encouragement of this book. I would also like to thank my father, Peter Tajmar, for his continued help towards publication, drawings and lots of advise.

This book is dedicated to my wife Yvonne and my son Alexander, who gave me the time to write this book during many nights in our kitchen!

Introduction

This book shall provide an up-to-date overview on a variety of advanced propulsion concepts, showing their limitations and potentials. Only necessary equations are given to understand the working principle without going into complicated details. The book is suitable for any technical student and everyone who is interested in how we will explore space in the future.

Chapter 1

Propulsion Fundamentals

1.1 History

The theoretical basis of all "classical" propulsion systems is the reaction principle published by Newton in his *Principia Mathematica* in the summer of 1687. It was the answer to astronomer Edmond Halley's question as to the nature of force that causes the movement and orbit of planets around the sun. This link to gravitation will be crucial when examining very advanced propulsion concepts in the last chapter of this book. However, rockets were built and flown well before Newton provided his mathematical principles. Feng Jishen, a Chinese who lived around 970 AD, is credited with the invention of the "Fire Arrow", a bamboo tube with a small hole in one end filled with gunpowder. It was used for amusement as a firework. The application as a weapon was soon realized and used against Japanese invaders in 1275. Mongolian and Arab troops brought the rocket to Europe and eventually to the United States.

In 1865, Jules Verne wrote his famous science fiction novel *Voyage from Earth to the Moon*. In this story a giant gun is used to send two men towards space. Contrary to rockets, a gun only provides an initial acceleration which is not practical since the required forces to reach space are far too powerful, no material could withstand them, even less a human being (nevertheless, an initial acceleration can reduce rocket propellant as we will see in chap. 3)! The idea of how to dramatically improve the performance of firework- and weapon-rockets, thus formulating the basic theory of spaceflight, independently occurred to three men on different continents around the beginning of the 20th century.

Konstantin Tsiolkovski (1857–1935) was a self-educated mathematics teacher living in Kaluga, a small city southwest of Moscow, Russia. In 1903, he published the article "The Investigation of Space by Means of Reactive Devices", describing in detail the connection between rocket fuel, velocities to reach space and the total mass of a spaceship. This relationship is still known as the Tsiolkovski equation (see section *Propulsion Fundamentals*). Furthermore he described liquid-fuelled rockets, artificial satellites and the concept of multi-stage rockets. His achievements were only theoretical, he never experimented with rockets himself. In the United States, Robert Goddard (1882–1945), a professor of physics at Clark University in Worcester, Massachusetts, was granted patents as early as 1914 on the design of

Konstantin Tsiolkovski (1857–1935)   Robert Goddard
(1882–1945, Courtesy of NASA)

liquid-fuelled rockets, combustion chambers, nozzles and even gyroscopes for guidance. He launched the very first liquid-fuelled rocket on the 16th of March 1926, a 5 kg rocket powered by liquid oxygen and petrol. In 1919, he published the article "A Method of Reaching Extreme Altitudes" in which he developed a precise theory of spaceflight including the designs and test results from his own experiments. He also wrote about the possibility of sending an unmanned rocket to the moon, which was ridiculed by the press and people at the time in general. For his further experiments he had to leave Massachusetts, moving to the desert in New Mexico where he continued his work until 1940. The third and maybe most influential spaceflight pioneer was Hermann Oberth (1894–1989), born in Siebenbürgen, at that time part of the Austro-Hungarian empire, and later moving to Germany. His doctoral thesis "The Rocket to the Planets" ("Die Rakete zu den Planetenräumen") was rejected at the University of Heidelberg and so he decided to publish it as a book. It became a best-seller, stimulating the amateur interest in rockets and leading to the foundation of many rocket societies in Germany, most important the *Verein für Raumschifffahrt,* in which Wernher von Braun was later a member. Many rocket enthusiasts appeared at that time such as Fritz von Opel and his rocket-powered car. This general excitement led to Fritz Lang, the famous director, making a movie entiteled "Woman in the Moon" ("Frau im Mond"). He commissioned Hermann Oberth in 1929 to build a model spaceship for the movie, and launch a rocket as a promotion-gag for the premiere. The movie company UFA financed experiments on Oberth's liquid-fueled rockets, and although he could not finish the rocket to be launched in time, key advancements on the technology were made. Moreover, the movie was a big success, resulting in even more interest in rockets and spaceflight.

Hermann Oberth
(1894–1989, Courtesy of NASA)

Wernher von Braun
(1912–1977, Courtesy of NASA)

The Treaty of Versailles that followed the end of World War I prohibited Germany from using long-range artillery. As a result, army leaders carried out research into rockets in order to circumvent this prohibition. As they had no experience in this field they turned to the amateur rocket societies, which were attempting to develop liquid-fueled rockets to send men into space. In 1932, Walter Dornberger contracted the young Wernher von Braun (he was already Oberth's assistant) to develop a long-range missile for the German army at the research station Peenemünde. The efforts of von Braun and his collaborators led to the A-4 rocket, better known as the V-2 (Vergeltungswaffe 2), which was the first medium-range missile. A much larger version, the A-9/10, was already on the drawing board before the end of World War II. It would have been the first intercontinental ballistic missile (ICBM) capable of delivering a nuclear warhead from Western France to New York. Both the Americans and the Russians raced towards Peenemünde both to capture key personnel as well as to take possession of any remaining V-2 rockets. Von Braun and most of his team immigrated to the United States where they continued their rocket research leading to the powerful Saturn-V rocket that brought the first men to the moon in 1969. The Russians (at that time Soviet Union) took most of Peenemünde's "treasures" and experimented first in Germany and later in Russia where Sergei Korolev and Valentin Glushko led the research. Korolev, as chief engineer, was the Russian counterpart to Wernher von Braun, and Glushko constructed the reliable rocket engines that are still used in part today. Although Russia put the first satellite (Sputnik) into orbit in 1957, sent several probes to the moon, and in 1961 put the first man in space (Yuri Gagarin), they lost the race to the

German V-2 Rocket   Russian   Buran and Energia
                           N-1 Rocket   (Courtesy of NPO Molniya)

moon. The reason that the powerful Russian N-1 moon rocket was never launched successfully is probably due to Korolev's premature death in 1966.

After the race to the moon, the United States developed the Space Shuttle orbiter (based on the original Russian concept BORAX), the first partly reusable spacecraft in the world, which was first launched in April 1981. Russia equalled with the Buran orbiter carried by the rocket Energia, but it only flew once due to the drastic reduction on space expenditure after the collapse of the Soviet Union. Technology demonstrators for the next generation launch vehicles are underway in the United States, that will just need refuelling like an aircraft. Other nations also operate launchers such as those of the highly successful Ariane program of the European Space Agency ESA, the Japanese H-Family, the Chinese Long March and the Indian PSLV and ASLV rockets. Smaller launchers have also been developed in Israel, Pakistan and Brazil.

New propulsion systems such as electric, propellantless or nuclear propulsion are required to fulfill today's mission requirements. In 1998, the US National Aeronautics and Space Administration (NASA) launched the Deep Space One spacecraft which accelerates Xenon ions to create thrust instead of using traditional liquid-fueled rocket engines, saving up to 80% of propellant. Soon spacecraft will

Saturn-V Rocket
(Courtesy of NASA)

Space Shuttle Orbiter (Courtesy of NASA)

operate on similar electric propulsion systems to fly to the moon and the planet Mercury. Manned missions to Mars will most probably rely on nuclear powered rockets, and yet there is another challenge to master: the first interstellar precursor spacecraft, which probably needs a completely new type of propulsion technology. NASA started its Breakthrough Propulsion Physics Program in 1996 to look at new concepts in propulsion to overcome current barriers in propellent, speed and energy. The future of propulsion technology from now on will be very different from the past.

1.2 Propulsion Fundamentals

As outlined in our historical overview, all classical propulsion systems simply rely on Newton's mechanics. From the reaction principle we know that every force acting upon a body will cause a reaction force of the same intensity in the opposite direction. This is the basis for the conservation of momentum which is illustrated in Fig. 1.1. A rocket expels part of its mass to produce thrust. However, one can also

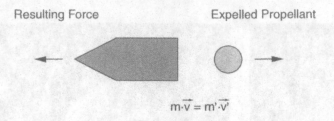

Fig. 1.1. Conservation of momentum

use external masses for propulsion purposes like solar wind particles or beam energy (e.g., using giant lasers) which are reflected/collected by huge sails, as we will see in later chapters. In that case, the total mass of a spaceship can be reduced drastically.

According to Newton, the force F produced by the propellant is the time derivative of its impulse:

$$\vec{F} = \frac{d\vec{I}}{dt} = \frac{d}{dt} m_p \cdot \vec{v}_p = \dot{m}_p \cdot \vec{v}_p. \tag{1.1}$$

We see that higher exhaust velocities v_p create the same force at a lower mass flow rate $\dot{m}_p$. This gives us information on how effective the propellant is used. Another characteristic parameter is the specific impulse I_{sp}, which is the propellant velocity divided by the standard gravitational acceleration $g_0 = 9.81\,\mathrm{ms^{-2}}$. This expression has traditionally been used for comparing propulsion systems with each other:

$$I_{sp} = \frac{v_p}{g_0}. \tag{1.2}$$

1.2.1 Tsiolkovski Equation

Let us assume a spacecraft of total mass m_0 (consisting of structure, payload and propellant) at a velocity v. If propellant is expelled, the velocity of the spacecraft will change accordingly. This can be expressed by equalling the force on the spacecraft and Eq. (1.1):

$$m_0 \cdot \frac{dv}{dt} = \frac{dm_p}{dt} \cdot v_p. \tag{1.3}$$

Since the structure mass of the rocket does not change, we can set $dm_p = dm_0$ and write:

$$dv = v_p \cdot \frac{dm_0}{m_0}. \tag{1.4}$$

Integrating the velocity for a mass change from m_0 to m

$$\int dv = v_p \cdot \int_{m_0}^{m} \frac{dm_0}{m_0}, \tag{1.5}$$

we obtain the famous rocket or Tsiolkovski equation:

$$\Delta v = v_p \cdot \ln \frac{m_0}{m}. \tag{1.6}$$

Using $m_p = (m_0 - m)$ we can transform Eq. (1.6) into

$$m_p = m_0 \cdot \left[1 - \exp\left(-\frac{\Delta v}{v_p} \right) \right]. \tag{1.7}$$

1.2.2 Delta-V Budget

All equations were derived assuming a constant propellant velocity v_p which is in almost all cases valid. The amount of propellant needed for a given mission is derived from the trajectory analysis. The corresponding propulsion system's performance is typically expressed as the required change of velocity from the spacecraft, Δv. The total requirement consists of several components,

$$\Delta v = \Delta v_g + \Delta v_{drag} + \Delta v_{orbit} - \Delta v_{initial}, \tag{1.8}$$

namely the Δv_g to overcome the gravitational potential (e.g., from the Earth's surface to the required orbit), Δv_{drag} due to drag from the atmosphere, Δv_{orbit} giving the velocity increment to reach a certain orbit and $\Delta v_{initial}$ the initial velocity. For example, if we launch from the Earth's surface and want to reach a Low Earth Orbit (LEO) at 100 km, the first factor is calculated by equating the difference of the gravitational potential energy between the Earth's surface and the 100 km orbit and the kinetic energy

$$\frac{mv^2}{2} = GMm \left(\frac{1}{r_{initial}} - \frac{1}{r_{final}} \right), \tag{1.9}$$

hence

$$\Delta v_g = \sqrt{2GM \left(\frac{1}{r_{initial}} - \frac{1}{r_{final}} \right)}, \tag{1.10}$$

where m is the mass of the launcher, M the Earth's mass, $r_{initial}$ and r_{final} are the radii at Earth's surface and at a LEO, respectively. By setting $r_{initial} = 6400\,km$ and $r_{orbit} = 6500\,km$, we obtain $\Delta v_g = 1.4\,km/s$. To stay in a LEO, the centrigufal force has to balance the gravitational pull

$$\frac{mv^2}{r} = \frac{GMm}{r^2},\tag{1.11}$$

where r is the radius from the center of the Earth to a LEO. We can now express the Δv_{orbit} requirement

$$\Delta v_{orbit} = \sqrt{\frac{GM}{r}}.\tag{1.12}$$

For a LEO say, $r = 6500\,km$, we get $\Delta v_{orbit} = 7.8\,km/s$, starting from the Earth's surface. Due to the Earth's rotation, $\Delta v_{initial} \simeq 0.4\,km/s$ (zero at the Earth's poles), the atmospheric drag adds another $\Delta v_{drag} = 0.1\,km/s$. The total requirement is therefore $\Delta v_{LEO} = (1.4 + 0.1 + 7.8 - 0.4)\,km/s = 8.9\,km/s$.

Once in a LEO, the Δv requirement for further targets such as planets is mostly influenced by the time required to reach the destination and its distance. For example, a robotic mission to Mars can last a year or more; for manned exploration, on the other hand, the trip time duration needs to be minimized to a few months only, which increases the Δv requirement significantly. Typical mission requirements are listed in Table 1.1. Of course, launching from a less massive planetary-type, such as the Moon or even in interplanetary space reduces the Δv requirement as well. Both maximum thrust and specific impulse for a variety of propulsion systems are listed in Table 1.2. Also a maximum Δv for a ratio of $m/m_0 = 0.1$ is listed using Tsiolkovski's equation (Eq. (1.6)). It can be clearly seen that not all propulsion systems are suited for all mission scenarios.

Table 1.1. Mission propulsion requirements (Frisbee, R.H., Leifer, S.D., AIAA 98-3403, 1998)

Mission	Description	Typical Δv [km/s]
LEO, GEO, Planetary Targets	Satellites, Deep space robotic missions	10–15
Human Planetary Exploration	Fast, direct trajectory	30–200
100–1,000 AU (Distance Sun–Earth)	Interstellar precursor mission	100
10,000 AU	Mission to Oort cloud	1,000
Slow Interstellar	4.5 light-years in 40 years	30,000
Fast Interstellar	4.5 light years in 10 years	120,000

Table 1.2. Typical propulsion performance

Propulsion System		Specific Impulse [s]	Maximum Δv [km/s]*	Maximum Thrust [N]
Chemical	Solid	250–310	5.7–7.1	10^7
	Liquid	300–500	6.9–11.5	10^7
Magnetohydrodynamic (MHD)		<200	4.6	10^5
Nuclear	Fisson	500–900	11.5–20.7	10^6
	Fusion	10,000–100,000	230–2,300	10^5
	Antimatter	60,000	1,381	10^2
Electric	Electrothermal	150–1,200	3.5–27.6	10^1
	Electrostatic	1,200–10,000	27.6–230	3×10^{-1}
	Eletromagnetic	700–5,000	16.1–115	10^2
Propellantless	Photon Rocket	"3×10^7"	unlimited	10^{-4}
Breakthrough		?	?	?

* Assuming $m/m_0 = 0.1$

1.2.3 Single-staging – Multi-staging

It is clear that the higher the exhaust propellant velocity v_p (or specific impulse I_{sp}), the less propellant needs to be carried onboard the spacecraft. Mass is usually directly linked to mission costs (1 kg payload costs around 20 k$ on the Space Shuttle and 5 k$ on a cheap Russian launcher). Chemical propulsion systems have exhaust velocities in the order of 3,000–4,000 m/s. Advanced propulsion concepts such as electric propulsion achieve velocities up to 100,000 m/s, which translates into huge propellant mass savings (and costs!). Figure 1.2 compares the mass ratio m/m_0 with the propellant velocity v_p for a typical launcher velocity of $\Delta v = 8,000$ m/s required to reach Low Earth Orbit at 100 km altitude. This mass ratio indicates the percentage of a launcher that is available for structure (tank, nozzles,...) and payload. Therefore it is also called *structural factor*.

Therefore, if one wants to reach orbit with a *single-stage rocket*, having an average $v_p = 3,500$ m/s, 90% of the rocket would be propellant and only 10% structure and payload. This is already a big challenge and new, very light materials together with rocket engines that have even better performance are currently under development in future launcher programs in the United States and Europe to test if it is possible with present day technology. Probably, if such a *Single-stage-to-orbit* (SSTO) launcher can be build, the payload capability would be practically zero.

Hence, a launcher is typically divided into several stages, each consisting of a separate engine and tank. If the first tank is empty, tank, engine and structure are dropped (probably with a parachute) and the overall launcher mass m_0 is reduced. This increases the achievable Δv – or the payload capacity for the same Δv. This concept is called *multi-staging* and was first described by Tsiolkovski in his 1924 article *Cosmic Rocket Trains*. He discovered the breakthrough that enables launchers to reach orbits.

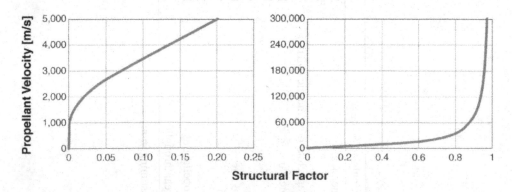

Fig. 1.2. Structural factor vs propellant velocity

For a two-stage rocket we have to divide the mass ratio m_0/m into two parts:

$$\left(\frac{m_0}{m}\right)_1 = \frac{m_{structure} + m_{propellant} + m_{payload}}{m_{structure} + (1-\alpha) \cdot m_{propellant} + m_{payload}}, \tag{1.13}$$

$$\left(\frac{m_0}{m}\right)_2 = \frac{(1-\alpha) \cdot [m_{structure} + m_{propellant}] + m_{payload}}{(1-\alpha) \cdot m_{structure} + m_{payload}}, \tag{1.14}$$

where α is the stage separator percentage of the first rocket part. We can calculate the gain Δv using Tsiolkovski's eq. (1.6):

$$\frac{\Delta v_{1+2}}{\Delta v} = v_p \cdot \frac{\ln\left(\frac{m_0}{m}\right)_1 + \left(\frac{m_0}{m}\right)_2}{\ln\frac{m_0}{m}}. \tag{1.15}$$

Figure 1.3 shows the velocity increment assuming an initial mass distribution for a 100 kg rocket of $m_{propellant} : m_{structure} : m_{payload} = 90{:}9{:}1$ kg. If the rocket is divided half-half, we get an increase of 1.23. Larger first stages result in larger velocity increments up to a maximum of 1.48 at $\alpha = 91\%$. More staging further increases rocket performance, but on the other hand increases also complexity and costs. For practical reasons, launchers with three stages are mostly used today. An example of the different stages in the Saturn-V moon rocket is shown in Table 1.3.

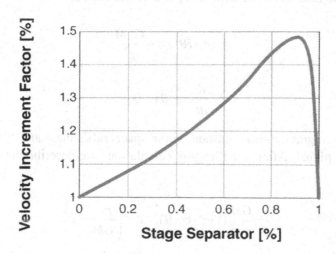

Fig. 1.3. Velocity increment factor vs stage separator for mass distribution propellant: structure: payload=90:9:1

Table 1.3. Saturn-V rocket stages – Payload to LEO orbit 118,000 kg

	Stage 1	Stage 2	Stage 3
Launch mass	2,286,217 kg	490,778 kg	119,900 kg
Dry mass	135,218 kg	39,048 kg	13,300 kg
Propellant	LO$_2$/Kerosene	LO$_2$/LH$_2$	LO$_2$/LH$_2$
Propellant velocity v_p	2,650 ms^{-1}	4,210 ms^{-1}	4,210 ms^{-1}
Mass ratio m/m_0	0.29	0.38	0.55
Velocity increment Δv	3,312 ms^{-1}	4,071 ms^{-1}	2,498 ms^{-1}

1.3 Trajectory and Orbits

All bodies in the universe follow orbits: the Earth, sun, stars, galaxies, etc. The planetary orbits around the sun were first explained by *Johannes Kepler* (1571–1630) interpreting the accurate observations of *Tycho Brahe* (1546–1601). The so-called Kepler laws can be derived from conservation of angular momentum and the hypothesis that all massive bodies attract each other directly proportional to $\frac{1}{r^2}$. We can use the same concepts to explain launcher and spacecraft dynamics. A rocket always delivers a spacecraft into an orbit where the gravitational pull by the Earth is counterbalanced by centrifugal forces. Spacecraft then either use their own propulsion system or a separate launcher upper-stage to reach their final destination. The shape of the orbit as we will see can only be controlled by the spacecraft's velocity v.

The balance between gravitational and centrifugal forces as well as the conservation of angular momentum can be written in polar coordinates as:

$$m \cdot (\ddot{r} - r\dot{\theta}^2) = -\frac{G \cdot M}{r^2}, \tag{1.16}$$

$$\frac{d}{dt}(mr^2\dot{\theta}) = 0, \tag{1.17}$$

where G is the gravitational constant, m the spacecraft's mass and M the mass of the orbiting planet. After some lengthy calculations we describe the orbital path by

$$\frac{1}{r} = \frac{GM}{r^2v^2}(1+\varepsilon \cdot \cos\theta), \quad \varepsilon = \left(\frac{rv^2}{GM}-1\right), \tag{1.18}$$

where ε is the eccentricity of the orbit. As shown in Fig. 1.4, the value of the eccentricity defines the shape of the orbit which can be a circle, ellipse, parabola

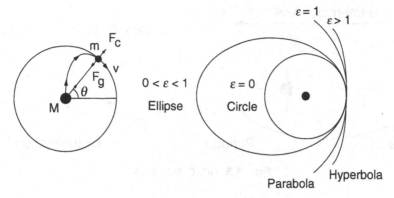

Fig. 1.4. Spacecraft orbits

or hyperbola. The parabola case $\varepsilon = 1$ defines the escape velocity, i.e. the spacecraft will just escape the planet's gravitational field and stop at infinity:

$$v = \sqrt{\frac{2GM}{r}}. \tag{1.19}$$

Due to the equivalent principle (inertial mass = gravitational mass), both escape velocity as well as the orbit trajectory do not depend on the spacecraft's mass! As an example, any body on the surface of the Earth ($M_{\mathrm{Earth}} = 5.98 \times 10^{24}\,\mathrm{kg}$, $r_{\mathrm{surface}} = 6370\,\mathrm{km}$) needs a velocity of $v = 11{,}190\,\mathrm{m/s}$ which is about 34 times the speed of sound (Mach 34) to escape the Earth's gravitational field. Note that this is only $\sqrt{2}$ times the velocity required to stay in a minimum orbit ($\varepsilon = 0$). Hence, a launcher needs to reach at least Mach 24 in order to insert a satellite into orbit! In addition to the mass requirements for a reusable SSTO, such high velocities induce substantial thermal stresses in the launcher structure. Up to now, high performance materials can only withstand such harsh environments for about 20 flights, afterwards they need to be replaced. Eccentricity values between 0 and 1 are typical orbit insertion velocities (Δv requirement for launcher), hyperbolic trajectories are used for interplanetary flight paths (i.e., mission to Mars).

1.3.1 Keplerian Orbital Elements in Elliptical Orbits

In order to fully specify orbit size, shape and orientation of an orbiting spacecraft or satellite (which is then of elliptical or circular shape), we shall define the so-called *Keplerian* (or classical) *orbital elements* as shown in Figs. 1.5 and 1.6:

(i) Semi-major axis a: Size of elliptical orbit.

(ii) Eccentricity e: Shape of orbit as described by Eq. (1.18). In case of an elliptical orbit the minimum distance between orbit and planet is called perigee whereas the maximum distance is called apogee.

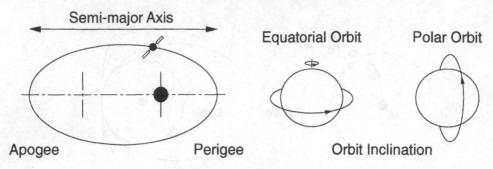

Fig. 1.5. Orbit elements

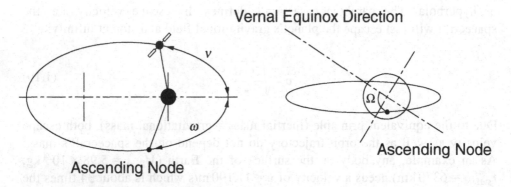

Fig. 1.6. Ascending node

(iii) Inclination i: Angle of orbit with the equatorial plane of the planet. An inclination of $i = 0°$ or $i = 180°$ is called equatorial orbit, $i = 90°$ is called polar orbit. We can further distinguish between a direct orbit ($0° < i < 90°$) and an indirect orbit ($90° < i < 180°$).

(iv) Longitude of ascending node Ω: In addition to the inclination i, the orbit can be inclined around its semi-major axis a. The ascending node is the point where the spacecraft passes the planet's equatorial plane from south to north. In case of an inclination around a, this point will be shifted. The longitude of the ascending node is defined as the angle between the vernal equinox direction and the ascending node.

(v) Argument of perigee ω: Angle between the ascending node and perigee. This defines the orientation of the orbit.

(vi) True anomaly υ: Angle between perigee and the spacecraft's location. This is the only Keplerian orbital element which changes with time.

1.3.2 Orbit Types

Several orbits are typical for space missions. Each of them need more or less Δv capability on the onboard propulsion system to maintain that orbit. Every orbit below 1000 km is called *Low Earth Orbit* (LEO). Space shuttles as well as the International Space Station ISS have altitudes of approximately 300 km but also many small satellites or mobile phone/data constellation satellites such as Iridium are common users of such an orbit. Comparing such altitudes to the radius of the Earth ($r_{surface} = 6370$ km) we see that they are not really far out in space. Communication satellites such as Astra TV broadcasting satellites need to rotate together with the Earth in order to remain at a fixed location (otherwise we would have to rotate our satellite dishes to follow the satellite!). This orbit is called *Geostationary Orbit* (GEO), has no inclination, is circular and has a radius of 42,120 km. The Russians were launching from a much higher latitude (e.g. from Baikonur) requiring a higher Δv than the US rockets in Cape Canaveral to reach GEO. Since the Russian rockets were not powerful enough in the 1960s to make up this difference only the United States had GEO geostationary satellites at that time. However, the Russians found a way by which their satellites could be used for communication purposes as well: a highly elliptical orbit of $e = 0.75$, $a = 26,600$ km and $i = 28.5°$ or $i = 57°$ has a perigee which does not rotate and remains in the northern (or southern depending on the chosen inclination) hemisphere for 11 hours until it moves very quickly to the opposite hemisphere in 1 hour, staying another 11 hours at apogee. Hence 11 hours per day could be used for communication – using three satellites in such an orbit a total coverage could be achieved. This orbit is known as *Molniya Orbit*, a ground track projection is shown in Fig. 1.7. Of course the optimum launch site for GEO orbit is close to the equator, such as the European Spaceport in French Guyana. A seabased launcher platform called *Sealaunch* has been developed to move to the best launch location for each specific mission. Another interesting orbit is the so-called *Sun Synchronous Orbit*. Here the spacecraft is always in direct orientation to the sun. All orbital types are summarized in Table 1.4.

Fig. 1.7. Molniya orbit ground track projection

Table 1.4. Orbit types

	Altitude	Orbital Parameters	Application
Low Earth Orbit (LEO)	<1000 km	–	Space Shuttle, Space Station, Small Sats
Geostationary Orbit (GEO)	42,120 km	$i = 0°$	Communication
Molniya Orbit	26,600 km	$e = 0.75, i = 28.5°/57°$	Communication, Intelligence
Sun Synchronous Orbit	700–7,300 km	$i = 95°$	Remote Sensing

1.3.3 Orbit Transfers

Every spacecraft needs to change at least one of the orbital parameters during its lifetime. There are several ways depending on the type of propulsion system and time constraints:

(i) Hohmann Transfer: This is the most common and fuel-efficient method of changing the orbit size (semi-major axis a). It needs powerful thrusters capable of delivering two Δv in a short amount of time. Imagine we want to move from one initial orbit to another one as shown in Fig. 1.8. The first burn increases (or decreases depending on whether we want to move to a higher or lower orbit) the spacecraft's speed and we move along an elliptical transfer orbit from perigee to apogee where we do a second burn in order to insert into the final orbit. A typical example is if a spacecraft needs to be inserted into GEO. Then the launcher first moves the spacecraft into the so-called *Geostationary Transfer Orbit* GTO (with apogee at 42,120 km), then another burn transfers the spacecraft into GEO. All velocity vectors are tangential to the orbits. Thus, the orbit changes only its size but not its plane.

(ii) Low Thrust Transfer Orbit: If the spacecraft has no powerful thruster on board such as a chemical or nuclear propulsion system, we can change to another orbit using a spiral transfer orbit as shown in Fig. 1.8. This of course requires much more time; however, as we will discuss in detail in the following chapters, such propulsion systems can save a big percentage of propellant needed. Such trajectory is typical for all kinds of electric propulsion systems, solar sails and tethers.

(iii) Gravity Assist Trajectories: If the Δv required for an orbit or plane change is too big, we can transfer angular momentum from a planet to the spacecraft gaining significantly in speed. This is a very common technique for almost all

Hohmann Transfer Orbit   Low Thrust Transfer Orbit

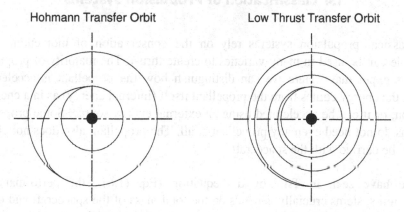

Fig. 1.8. Hohmann vs Low Thrust Transfer Orbit

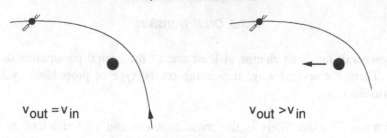

$v_{out} = v_{in}$ $v_{out} > v_{in}$

Fig. 1.9. Gravity assist trajectory

interplanetary missions. If a spacecraft is approaching a planet, the gravitational field attracts the spacecraft, and it speeds up. After passing the planet, the gravitational field will slow down the satellite. Hence from the planet's point of reference the spacecraft did not gain any net velocity. But the planet is at the same time moving along its orbit around the sun. In this constellation, the planet pulls the spacecraft gravitationally with it along its path. This causes a net velocity gain from the spacecraft (and a net orbital velocity decrease of the planet) from the sun's point of reference (Fig. 1.9). The more massive the planet, the better the effect. That's why all major gravity assists are done with Jupiter as the most massive planet in our solar system (one needs to pass the planet very closely – this is not possible with the sun!). Since momentum is the product of mass and velocity, the velocity change for the spacecraft is significant and the velocity change for the planet truly negligible.

(iv) Aerobrake Trajectory: If a spacecraft is moving close to a planet's atmosphere, it can be slowed down due to drag. With this simple technique it is possible to be captured into a planetary orbit without consuming expensive onboard propellant.

1.4 Classification of Propulsion Systems

All classical propulsion systems rely on the conservation of momentum. This principle can be used in many varieties to create thrust. The majority of propulsion systems expel mass. Here we can distinguish how the propellant is accelerated. Either, the energy comes from the propellant itself (internal energy) as in a chemical reaction, or it can be accelerated using an external energy source. Some propulsion systems do not need even any propellant at all. The propellant also does not always need to be carried with the spacecraft.

As we have seen in Tsiolkovski's equation (Eq. (1.6)), the performance of propulsion systems crucially depends on the total mass of the spacecraft and on the speed of the propellant. Hence all propulsion systems which reduce the need of

Table 1.5. Classification of propulsion systems

	Internal Energy	External Energy	External/Internal Energy
Internal Propellant	Chemical	Electric	Nuclear (Induction Heating)
	Nuclear (Fusion, Fission, Antimatter)		
	Air Breathing		
External Propellant	MHD	Propellantless (Laser, Solar Sail)	Air Breathing
			Propellantless (Tethers)
External/Internal Propellant		Propellantless (Solar Sail)	
No Propellant	Propellantless (Photon, Nuclear)	Catapults	Breakthrough Propulsion

propellant making the spacecraft lighter are considered as advanced propulsion systems which is the topic of this book. Also propulsion systems which increase the speed of the propellant fall under this category. Therefore, we can classify all propulsion systems in terms of energy and mass usage as summarized in Table 1.5 defining the chapters of this book.

All propulsion systems are chosen to meet a certain mission's Δv and maximum thrust requirement. These requirements come from the trajectory analysis. Many factors like gravity assists, initial transfer orbit, spacecraft mass, but also the time to reach the destination influence these factors. For instance, a manned mission to Mars will require a fast transfer time in the order of a few months and thus a very direct trajectory and high Δv and thrust. On the other hand, robotic missions may use a couple of years involving many gravity assists and a much less powerful propulsion system. Typical mission propulsion requirements are shown in Table 1.1. Both maximum thrust and specific impulse, necessary to calculate the Δv using Tsiolkovski's equation (Eq. (1.6)), for a variety of propulsion systems are listed in Table 1.2.

Chapter 2

Chemical Propulsion Systems

This chapter deals with classical rockets – and how they can be improved. A rocket is basically a thermodynamic system: it converts heat, generated by burning propellants, into kinetic energy through a nozzle. The energy needed for accelerating the propellant is provided by the propellant itself. This is done through a chemical reaction between a fuel and oxidizer. These can be solid or liquid, there are also monopropellant engines which use a catalyzer to decompose and start the reaction. Chemical rocket engines exist for a wide range of thrusts, from 0.01 N up to 10^7 N. They are the only ones up to now that have been powerful enough to lift a spacecraft from the Earth's surface (launcher – primary propulsion system).

2.1 Thermodynamic Characterization

The energy released by the temperature gradient between the combustion chamber T_c and the nozzle exit T_e can be described by assuming a perfect gas:

$$E_{\text{gas}} = c_p \cdot m_p \cdot (T_c - T_e),\qquad(2.1)$$

where c_p is the specific heat at a constant pressure. We can then calculate the propellant velocity v_p by equaling this energy with the kinetic energy $\frac{1}{2}m_p v_p^2$

$$v_p = \sqrt{2 \cdot c_p \cdot (T_c - T_e)}.\qquad(2.2)$$

In classical gas theory, a reversible thermodynamic process relates temperature and pressure with the ratio of specific heat γ at constant pressure to that at constant volume

$$T \cdot p^{\gamma/(\gamma-1)} = \text{constant}.\qquad(2.3)$$

The specific heat ratio γ is 1.3 in air at normal temperature conditions. In case of high temperature gases such as rocket exhaust plumes, a typical value is 1.2. Together with the universal gas constant R and the molecular weight of the exhaust gas m_{gas} we can rewrite the specific heat as

$$c_p = \frac{\gamma}{(\gamma - 1)} \cdot \frac{R}{m_{gas}}.$$ (2.4)

Finally we substitute and express the propellant velocity as

$$v_p = \sqrt{\frac{2\gamma}{(\gamma - 1)} \cdot \frac{R \cdot T_c}{m_{gas}} \cdot \left[1 - \left(\frac{p_e}{p_c}\right)^{(\gamma-1)/\gamma}\right]}.$$ (2.5)

This equation provides us with some insight into where chemical propulsion systems can be optimized. For instance, the ratio p_e/p_c is defined by the nozzle design and is also influenced by the ambient atmospheric pressure. In this case, the specific impulse will be at maximum when the rocket is fired in vacuum where $p_e = 0$. Another important parameter is the temperature in the combustion chamber T_c which is a function of the energy released by the chemical reaction of the propellants. Research on high energy density propellants is being conducted to improve this value. Lightweight chamber material that can withstand such high temperature as well as the mechanical stresses during launch are another area of improvement. The molecular mass m_{gas}, however, is a compromise which has to be chosen: if it is low, the specific impulse will be higher; on the other hand, a low molecular mass also reduces the mass flow which is directly proportional to the thrust (Eq. (1.1)).

2.2 Chemical Propulsion Overview

Referring to the physical state of the stored propellant they can be divided into three different categories:

2.2.1 Liquid Propulsion Systems

The liquid propellant is stored in tanks and fed on demand into a combustion chamber. This can be done by either only gas pressurization (this is needed anyway to empty the whole propellant tank) or a pump. Since high temperatures are required for high specific impulses, cooling of the combustion chamber and the nozzle is very important. Sometimes, the propellant is also stored in a gaseous state. However, since the gaseous state requires much more volume (and mass!) than the liquid state, this is only possible for small engines. Research is even being carried out to further reduce the volume by using a mixture of solid and liquid fuel – the so-called *slush*.

Monopropellant Engines

Monopropellant engines are most widely used for spacecraft attitude and orbit control. Typical propellants are hydrazine (N_2H_4) or hydrogen peroxide (H_2O_2) due

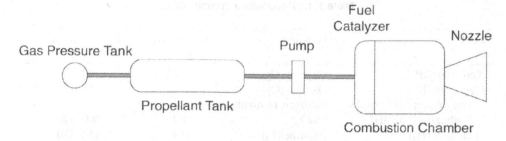

Fig. 2.1. Monopropellant engine

to their excellent handling, throttling possibilities and relative simplicity (Fig. 2.1). Due to the catalyzer needed before the combustion chamber, the operative pressure is not very high. This translates into rather low specific impulses of 150–250 s.

Bipropellant Engines

In this design, the fuel and oxidizer are split and fed separately into the combustion chamber (Fig. 2.2). The stability of the feeding process as well as maintaining an optimal mix of the two propellants are very critical issues. Depending on the propellant, a separate plug is needed for ignition, or the propellants are hypergolic and ignite at contact (as in the Space Shuttle Main Engine). There are a large variety of fuel combinations available with specific impulses ranging from 200–470 s. Some important propellants are listed in Table 2.1. Not all combinations are really practical. For instance although fluorine and hydrogen produce the highest specific impulse, fluorine is highly corrosive. Other propellants are toxic. The best compromise has to be selected from case to case according to mission requirements. The largest bipropellant engines produced so far are the F1 main engine from the Saturn-V launcher (Fig. 2.3) and the Russian counterpart RD-170 from the Energia launcher. The main disadvantage of bipropellant engines are their complexity and cost.

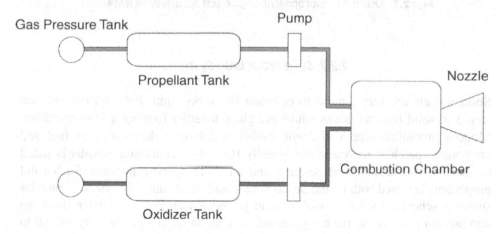

Fig. 2.2. Bipropellant engine

Table 2.1. Bi-propellant combinations

Fuel	Oxidizer	Average density [g/cm^3]	Specific impulse [s]
Kerosine (RP-1)	Oxygen (O$_2$)	1.02	300–360
Hydrogen (H$_2$)	Oxygen (O$_2$)	0.35	415–470
Unsymmetrical Dimethyl Hydrazine (UDMH)	Nitrogen Tetroxide (N$_2$O$_4$)	1.20	300–340
Hydrogen (H$_2$)	Fluorine (F$_2$)	0.42	450–480

Fig. 2.3. Saturn V-F1 bipropellant engine test (Courtesy of NASA)

2.2.2 Solid Propulsion Systems

Solid rockets are very similar to common fireworks. Both fuel and oxidizer are stored in solid form as grains which are glued together forming a kind of rubber. Modern propellants are a synthetic rubber based on hydrocarbons as fuel and ammonium perchlorate as oxidizer. Usually 16–18% of aluminium powder is added to increase the chamber temperature and therefore specific impulse. Such solid propellants are used both on the Space Shuttle and the Ariane 5 boosters. Figure 2.4 shows a schematic setup of such a solid propulsion engine. An igniter starts the combustion process and the hot gases are expelled through a nozzle very similar to

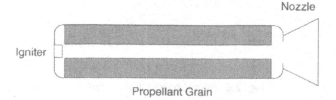

Fig. 2.4. Solid propulsion engine

Fig. 2.5. Solid propellant grain types

liquid propulsion systems. The solid grains can be arranged in many different shapes influencing the thrust profile. Figure 2.5 shows several common types. The progressive type burns from the inside to the outside surface increasing both mass flow rate and thrust with time. The surface area and therefore thrust is much more constant in the star type of shape. The flattest thrust profile is achieved in the ring shape, which is more complicated to manufacture. As the main disadvantage, solid rockets cannot be stopped easily once they are fired. Some methods include the injection of a special liquid or special window openings to rapidly reduce the inside pressure. Also the specific impulse is not very high (260–310 s) due to the relatively high molecular weight of the propellant. However, solid rocket motors are simple, reliable (absence of complex turbo pumps, separate propellant tanks,...) and cheap. Moreover, the high mass flow rate can create very high thrust levels up to 10^7 N which make them ideal for supporting launchers during the early phase (as in the Space Shuttle or Ariane 5).

2.2.3 Hybrid Propulsion Systems

Hybrid rockets try to surpass the shutdown and restart problem of solid rocket engines by storing either fuel or oxidizer in liquid and the other one in solid form (Fig. 2.6). This increases safety and also enables throttling of the engine. Mostly hydroxyl-terminated polybutadine as solid fuel grain and liquid oxygen as oxidizer is used nowadays. The specific impulse in this case is between 240–270 s. Although studies on this type of rocket go back to space pioneer Hermann Oberth, several technical hurdles remain and must be overcome before reliable hybrid rockets are feasible.

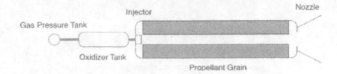

Fig. 2.6. Hybrid propulsion engine

2.3 Nozzle Design – Atmospheric Flight

Up to now we have neglected the influence of the atmosphere and the nozzle on the performance of a rocket. Expansion of hot gas without a nozzle, just through a hole, also produces a thrust, but it is about 50% less than that produced by using an optimally performing nozzle in vacuum.

2.3.1 Cone–bell Shaped Nozzle

The most simple design for a nozzle is the cone shape. More common is the bell shape (Fig. 2.8), which reduces the plume divergence after the nozzle exit. If the rocket engine is operated in atmosphere, the outside pressure is balanced everywhere on the surface of the rocket except on the nozzle exit. Here we have a pressure balance between the exit pressure from the exhaust hot gases p_e and the atmosphere pressure p_a. If both pressures are not equal an additional force appears in our momentum balance from Eq. (1.1), depending on the pressure difference and the nozzle exit area A_e.

$$F_{\text{axial}} = \dot{m}_p \cdot v_p + (p_e - p_a) \cdot A_e \qquad (2.6)$$

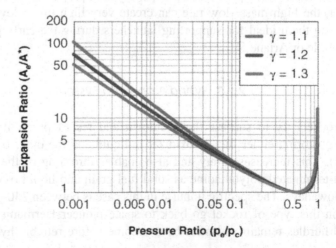

Fig. 2.7. Expansion ratio vs pressure ratio

Fig. 2.8. Cone-bell shaped nozzle

The exit pressure can be controlled by the length of the nozzle and the throat to exit area ratio A_e/A^* (*expansion ratio*) of the nozzle; a longer nozzle and a high expansion ratio reduce the exit pressure. The relation between the expansion and pressure ratio is shown in Fig. 2.7.

However, as we have already seen in our thermodynamic derivation in Eq. (2.5), the propellant velocity is also affected by the exit pressure. In fact it turns out that for a fixed ratio between p_c and p_a the maximum thrust is produced if $p_e = p_a$ which is called *optimum expansion*. Accordingly $p_e > p_a$ is called *under-expansion* and $p_e < p_a$ *over-expansion*. As already pointed out, the maximum thrust will be produced in vacuum where $p_a = 0$.

If the rocket operates at sea-level altitude at the beginning of a typical launch campaign, p_e needs to be high in order to compensate the ambient air pressure. Hence the nozzle shall be short and the expansion ratio low (typical value is 10). As the rocket is moving towards higher altitudes the atmosphere gets thinner and p_a smaller. For optimal performance, the expansion ratio or nozzle length should adapt accordingly – otherwise the nozzle is only optimized for a very specific altitude. Usually, as a launcher consists of several stages, each stage has a different expansion ratio in order to produce the optimum thrust in its atmosphere environment (third stages have expansion ratios in the order of 80). There are also concepts of how to modify the expansion ratio or the nozzle length during flight as shown in Fig. 2.9. However, these concepts are not very reliable since the release of a droppable insert or the correct mounting of an extension on the nozzle is a very complicated task during rocket operation. For operation in vacuum ($p_a = 0$), the ideal rocket nozzle should be infinite long and the expansion ratio infinite high.

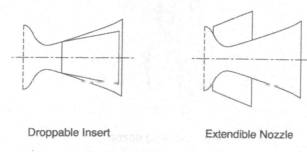

Droppable Insert   Extendible Nozzle

Fig. 2.9. Two-step nozzles

A bigger nozzle also increases the mass and cost budget. Hence there is a certain trade-off between better performance and design limitations.

2.3.2 Advanced Nozzles with Aerodynamic Boundaries

A nozzle shapes the expelled hot gases and increases the resulting thrust. The nozzles discussed so far shape the gases with an outer boundary. Another approach is to shape the gases against a surface inside the plume and let the ambient atmosphere act as the outer boundary. In this configuration, the beam automatically adopts its expansion ratio (if the atmosphere gets thinner the plume will expand more and the expansion area will increase) and operates at performance close to maximum independent of the altitude.

Plug Nozzle

Here, the exhaust gases are injected along an annular aperture between the combustion chamber and a conical plug (Fig. 2.10). It then flows along the plug and expands according to ambient atmospheric pressures. For ideal performance, the plug should have the same length as an optimally performing conical or bell-shaped nozzle. Although the advantage of altitude independent performance is significant, several problems were identified in this concept. The most important difficulty is how to cool the plug adequately while it is exposed to hot gases. Also annular combustion chambers are much more complicated to construct than simple cylindrical ones.

Aerospike Nozzle

The plug nozzle problems can largely be overcome by truncating the plug and replacing the missing part with the injection of cool gases. Moreover in a variation

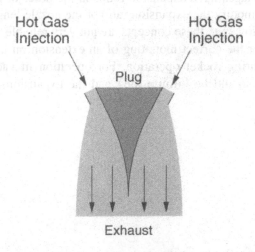

Fig. 2.10. Plug nozzle

called linear aerospike nozzle, the cylindrical geometry is transformed to linear geometry solving the problem of annular combustion chambers. The smaller plug also saves mass and eases vehicle/engine integration.

2.4 Advanced Propellants

Only small changes in the propellant density as well as in the specific impulse of an engine can significantly increase the payload capability. For example if we consider a SSTO vehicle using a LO_2/LH_2 bipropellant engine, an increase of 10% in payload density (e.g., due to the use of a hydrogen slush – a mixture of solid and liquid hydrogen) would result in a payload increase of more than 25%. Because Tsiolkovski's rocket equation is an exponential law, an increase of 10% in specific impulse results in an even more significant increase of 70% in payload capability.

2.4.1 Tripropellants

Many chemical reactions provide a larger specific energy release than LO_2/LH_2 but are unacceptable rocket propellants because their reaction product, or a significant part of it, is not gaseous. One way of approaching this is to use hydrogen as a working fluid in addition to fuel and oxidizer in bipropellant liquid propulsion systems.

Attractive propellant combinations are Beryllium – Oxygen (Be/O_2) or Lithium–Fluorine (Li/F_2). The metals can be either melted and used as normal

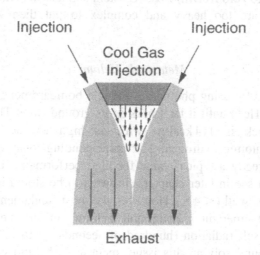

Fig. 2.11. Aerospike nozzle

liquid propellant or injected in a stream of the oxidizer or hydrogen. The specific impulse of such tripropellant rocket engines is around 700 s which is about 55% higher than for LO_2/LH_2 bipropellant systems. However, there are some practical problems, namely a bad combustion efficiency around 85%. Also the exhaust gases are toxic and can cause substantial contamination problems.

2.4.2 High Energy Density Matter (HEDM)

High Energy Density Matter (HEDM) are atomic or metastable propellants, which have a very high recombination or return-to-ground state energy. Those exotic propellants can be added to normal propellants significantly increasing the specific impulse. Obviously, storage of this HEDM is one of the major difficulties. Most of them can be trapped in a solid matrix at very low temperatures. For example, such a matrix can be combined with solid propellants either in solid propulsion systems or in hybrid rockets where the solid part is doped with HEDM species.

Atomic Hydrogen

Atomic hydrogen (H), formed by dissociating diatomic hydrogen molecules (H_2), stores 52.2 kcal/g, which is released during recombination (H–H). This is more than one order of magnitude than the energy release of 3.2 kcal/g from H_2/O_2. The ideal specific impulse is 2,112 s. Storage is a major problem because the solid matrix can only trap 1.5 percentage of its weight with atomic hydrogen. Additionally, high magnetic fields are required to orient the atomic spins in such a way as to lower recombination probabilities inside the matrix. Some research has been done into the storage of atomic hydrogen only inside a magnetic bottle or in a extremely low temperature (close to zero Kelvin) Bose–Einstein condensate. However, all methods mentioned so far are too heavy and complex to put them on a launcher or spacecraft.

Metastable Helium

If helium is excited by using photons or particle bombardment, it can become so called metastable (He*) until it falls back in its ground state. The energy released during this fallback is 114 kcal/g – twice as high as the one obtained by recombination of atomic hydrogen. The corresponding ideal specific impulse is 3,150 s. This is already a typical value for high performance electric propulsion systems as we will see in later chapters. It needs to be stored in a matrix at very low temperatures as well (< 4 K). However, the most fundamental problem is that even at such low temperatures, metastable helium can only be stored for a very short time due to self-radiation (hundred of seconds up to 2.3 hours). Unless a breakthrough is found solving this issue, metastable helium cannot be used for propulsion purposes.

Metallic Hydrogen

The metallic state of hydrogen is formed when solid hydrogen is compressed under very high pressure (about 1.4 Mbar). When released to normal nonmetallic conditions, substantial energy is released with an estimated specific impulse of 1,700 s. Another attractive feature is the high propellant density of 1.15 g/cm³ compared to 0.088 g/cm³ for solid hydrogen. Research on metallic hydrogen is still at a very early stage, mass production and storage is still unclear and needs further research.

2.5 Alternative Designs

2.5.1 Pulse Detonation Rockets

In a pulse detonation rocket, combustion occurs at a constant volume instead of at constant pressure. This increases the thermodynamic efficiency by 10% and leads (theoretically) to a higher specific impulse. The combustion chamber has the shape of a tube with one closed end and the other able to be opened by a valve. Sufficient propellant is fed into the tube and is then ignited near the closed end. A compression wave travels through the tube at the speed of sound and ensures complete combustion before the hot gas is expended. The valve opens, the hot gas is released, the valve is closed and the cycle starts again. Since the propellant is injected at a lower pressure than in conventional chemical rocket engines (they must inject it at a higher pressure than the chamber pressure to ensure a continuous combustion), the pumps can be much lighter and cheaper. Also the overall geometry of such pulse detonation rockets appears to be more simple. The most important difficulty of these engines is to produce a constant thrust out of the pulses. Pulse detonation jet engines (similar to actual rocket engines) were used for the German V-1 flying bomb during World War II.

2.5.2 Rocket-based Combined Cycle (RBCC)

As we have seen, the propellant can take up to 90% of the rocket's mass. If the engine were to use oxygen from the air during ascent instead of carrying it onboard, a drastic increase in payload capability could be achieved. This is the aim of a rocket-based combined-cycle (RBCC) engine. Basically, it is a combination of a classical rocket and jet engine inside an open tube (Fig. 2.12) Their operation modes, depending on the velocity, can be subdivided as follows:

(i) In the ejector mode, the rocket engine works like the compression stage for the jet engine. The high velocity rocket exhaust gas entrains and compresses air flowing from the open end of the tube.

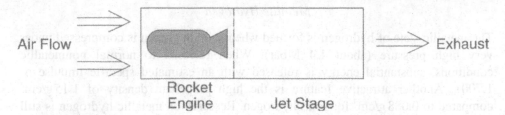

Air Flow

Exhaust

Rocket
Engine   Jet Stage

Fig. 2.12. Rocket-based combined-cycle (RBCC) engine

(ii) In the ram jet mode, the rocket engine is turned off when the vehicle velocity reaches about Mach 2. At this point, the ram pressure is enough to compress the inlet air. Fuel is added and burned in the jet stage.

(iii) In the scram jet mode, the secondary fuel injection from the jet stage is moved forward to allow adequate time for fuel and air to mix before burning to provide the thrust.

(iv) The final pure rocket mode begins when the vehicle reaches such a high altitude that the no longer ambient air is available.

This design offers significant advantages, the much increased payload fraction lowers costs and weight. Also reliability and safety can be greatly improved. The most difficult tasks in this concept are supersonic combustion required for ram and especially the scram jet modes. Russia tested scram jet technologies on their missiles in the past. Active research of this rocket engine concept is being carried out by NASA, Boeing and Aerojet in the United States.

2.5.3 Rotary Rockets

Safety and reliability are key issues for advanced launcher technologies, especially if they are to be capable of human transportation. The part of a rocket engine under constant high mechanical stress is the turbo pump injecting propellant into the combustion chamber at very high speeds. This can be outpassed by a very simple solution, the rotary rocket. In this concept, the combustion chamber and nozzle are mounted on rotors (similar to helicopters) and the propellant tank is mounted in the middle. Radial propellant feed lines connect the chamber with the tank. If the rotor rotates, the propellant is pressed into the chamber by centrifugal forces alone avoiding turbo pumps completely! Only low rotation speeds are necessary to achieve sufficient pressure. The rotor in fact can also operate as a helicopter enabling soft landing. Such a vehicle was under construction at Rotary Rocket in the United States (Fig. 2.13), but unfortunately further development stopped. Three test flights with an early prototype were conducted in 1999.

Fig. 2.13. Rotary rocket (Courtesy of Rotary Rocket)

2.6 Reusable Launch Vehicles

All these new propulsion technologies are to be implemented in a new generation of launch vehicles providing

(i) reduction in launch costs: at least one order or magnitude from ten thousands of $/kg to a few hundred $/kg,

(ii) increased safety: current launchers fail between 1–10% – reduction down to 0.1% or better,

(iii) increased reliability: fully reusable parts and routine operations will lead to an improvement by one order of magnitude at much lower costs

Spaceflight will become as simple as flying airplanes. Several countries are working in this direction, their achievements are summarized below. Even private companies are working on the development of such launchers (e.g., Kistler, Rotary Rocket,...). The most desirable concept is of course a fully operational SSTO launcher. However, also fully reusable TSTO concepts are being studied since their technology is closer to realization.

2.6.1 United States Future Launchers

NASA's Advanced Space Transportation Program (ASTP) was initiated in 1994 with the signing of President Clinton's National Space Transportation Policy. NASA was assigned the lead for the development of reusable launch vehicle (RLV) technologies that would pave the way for future low-cost systems, capable of flying for one-tenth the cost of today's systems.

NASA is operating the Space Shuttle fleet which is already a 1st generation RLV–the main orbiter is flying back, landing like an airplane, and being reused after refurbishing. The solid boosters were intended for reuse first, but a proper refill and refurbishment turned out to be more expensive than a complete new one. Liquid propellant boosters, though initially more expensive than solid ones, could be reused more easily and improve performance (higher I_{sp}). Such an option is currently under study (the Russian Buran-Energia orbiter featured liquid boosters from the beginning!). The tank, however, can not be reused at all since it burns out on reentering the atmosphere. True fully reusable 2nd generation RLV are currently under construction, and some concepts have already been tested. The United States is continuously building and operating experimental research vehicles (X-planes), flight testing advanced technologies.

X-15

The first one directed to reusable launcher technology was the North American X-15 rocket plane, which was first already flown back in 1959 (Figs. 2.14 and 2.15). It set the world record both in speed (Mach 6.72) and altitude (108 km) for winged aircraft, which remain unsurpassed. A scram jet engine should have been tested as well but was canceled and the program stopped in 1968. It was a high-speed

Fig. 2.14. X-15 (Courtesy of NASA)

Fig. 2.15. X-15A2 with external fuel tanks (Courtesy of NASA)

research aircraft used to provide information on thermal heating, high speed control and stability, and atmospheric re-entry. The X-15 was dropped from a B-52 bomber and then propelled by a Thiokol XLR99-RM-2 throttleable liquid fuel (liquid hydrogen, anhydrous ammonia) rocket.

DC-XA Delta Clipper

The first vertical takeoff and vertical landing (VTOVL) SSTO prototype oriented to sub-orbital flight was the McDonnell Douglas Delta Clipper Experimental (DC-X). Constructed in 1991–1993, it was a one-third scale model of a real SSTO launcher (Figs. 2.16, 2.17). The DC-X featured four Pratt and Whitney RL-10A-5 liquid propellant rocket engines (LO_2/LH_2), had a total mass of 16,320 kg, a diameter of 3.1 m and a length of 11.4 m. It flew eight times till 1995 when it was converted

Fig. 2.16. Delta clipper DC-XA at launch (Courtesy of NASA)

Fig. 2.17. Delta clipper DC-XA during flight (Courtesy of NASA)

into the DC-XA (Delta Clipper Experimental Advanced) model with a lightweight graphite-epoxy liquid hydrogen tank, an advanced graphite–aluminum honeycomb intertank and an aluminum–lithium liquid oxygen tank leading to a mass reduction of 620 kg. The longest flight had a duration of 140 s at an altitude of 1.25 km in 1996. Unfortunately, the prototype was destroyed after the 4th flight as the vehicle tipped over during landing due to a broken landing strut and the LOX tank exploded.

X-33

Another subscale SSTO prototype was under development by Lookheed Martin Skunk Works from 1996–2001 under the X-33 program (Fig. 2.18). It is based on a lifting body shape with two linear aerospike rocket engines (Fig. 2.19) and a rugged metallic thermal protection system. Moreover, it also features lightweight components and a composite hydrogen and aluminium oxygen fuel tank built to conform to the vehicle's outer shape. The X-33 was conceived as an unpiloted vehicle, taking off vertically like a rocket, reaching an altitude of up to 100 km, speeds faster than Mach 13, and landing horizontally like an airplane. The turnaround time between X-33 test flights was aimed at only seven days, during the initial test-phase even two days. The main characteristics are listed in Table 2.2. First test flights were scheduled for 2002 after some problems with the lightweight tanks. Currently, the program is on hold. If successful, Lookheed Martin intends to proceed with the full-scale SSTO model named Venture Star.

X-34

The X-34 was intended as a reusable technology testbed vehicle for a TSTO concept from Orbital Sciences Corporation. The vehicle structure is an all-

Fig. 2.18. X-33 and venture star (Courtesy of NASA)

Fig. 2.19. XRS-2200 linear aerospike engine (Courtesy of NASA)

composite with a one piece delta wing design, 17.7 m in length and 8.5 m wide. The
X-34 vehicle was to be powered by a LOX/Kerosene liquid MC-1 engine capable of
speeds of up to Mach 8 and altitudes of 76 km. It should have been launched from
the Orbital's L-1011 aircraft, the same that carries the Pegasus booster (Fig. 2.21).
Specific technologies that were designed for the vehicle are, for example, composite
structures, a composite kerosene fuel tank, an advanced thermal protection system,
leading edge tiles and autonomous flight operations. One of the main project goals
was also to demonstrate low cost flight operations ($500K/flight recurring cost) and

Table 2.2. X-33 Specifications

Length	21 m
Width	23.5 m
Take-off Weight	142,500 kg
Fuel	LO_2/LH_2
Fuel Weight	105,000 kg
Main Propulsion	2 XRS-2200 Linear Aerospike Engines
Maximum Speed	Mach 13+

Fig. 2.20. X-34 Technology demonstrator (Courtesy of NASA)

a rate of 24 flights in 12 months while maintaining a small work force. The project was canceled mainly due to funding reasons in 2001 (Fig. 2.20).

X-43 (Hyper-X)

This program is to demonstrate Scramjet technology. Three research aircraft are to fly up to Mach 10, beating the early record of the X-15 s. Hyper-X is to ride on a Pegasus-derived booster rocket (Fig. 2.21), built by the Orbital Sciences Corp., which will be launched by a B-52 bomber from an altitude of 5,800 m to 13,100 km, depending upon the mission. For each flight, the booster will accelerate the Hyper-X research vehicle to the test conditions (Mach 7 or 10) at approximately 30,000 m, where it will separate from the booster and fly under its own power and preprogrammed control. During the first test flight in 2001, a major malfunction occurred and the prototype was lost. After an extensive review, two additional flights are planned.

Fig. 2.21. Hyper-X launched with Pegasus booster (Courtesy of NASA)

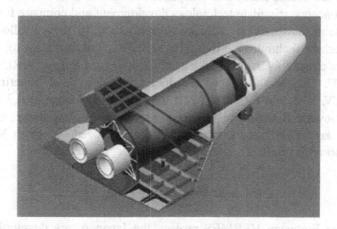

Fig. 2.22. Proposed European future launcher (Courtesy of ESA)

2.6.2 European Future Launchers

A 1st generation RLV similar to the US Space Shuttle was under advanced development by the European Space Agency ESA. Named HERMES, it was intended to be attached on top of the Ariane 5 rocket. The projected high costs stopped the program in the mid 1990s. One of the first proposed 2nd RLV was the German TSTO concept Sänger and later the Sänger II in the late 1980s. It consisted of a hypersonic Mach 4 aircraft booster equipped with ramjets as the first stage and a rocket powered second stage named Horus. Unfortunately the funding was cut although initial ramjet engine tests were successful. ESA started its Future

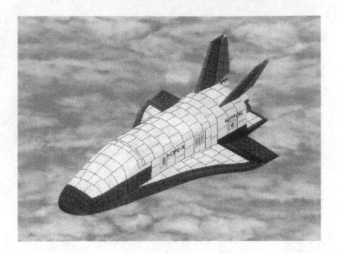

Fig. 2.23. HOPE-X technology demonstrator (Courtesy of NASA)

European Space Transportation Investigations Programme (FESTIP) in 1994, to study RLV concepts, stimulate technology development and compare fully reusable to partly reusable launchers. It continued work till 2000. In 1999, a follow-up called Future Launcher Technology Programme (FLTP) was initiated to pursue the developments already made and make a decision about a European RLV prototype in 2006–2007. One of the proposed vehicles, the European Experimental Test Vehicle EXTV, is shown in Fig. 2.22. It has a projected mass of 4,2 tons and is planned for several flights. The vehicle is to be powered by rocket engines and planned to take off and land horizontally, reaching speeds of up to Mach 4. An additional booster is planned to increase the flight regime to Mach 12.

2.6.3 Japanese Future Launchers

Similar to the European HERMES project, the Japanese are developing a small shuttle planned to fly on top of their H-II heavy launcher. The originally proposed HOPE (H-II Orbiting Plane) was supposed to be a major logistical vehicle for the Japanese Experiment Module of the International Space Station ISS. Budget cuts led to a shift towards a scaled demonstration vehicle called HOPE-X (Fig. 2.23) which shall also serve as a testbed for future RLV technology. High Speed Flight Demonstrator experiments were scheduled for 2002 to 2004, but another budget cut has put this project currently on hold as well.

Chapter 3

Launch Assist Technologies

As we have seen, a completely reusable launch vehicle with a good payload capacity – preferably single-stage-to-orbit – is very difficult to achieve with purely onboard chemical propulsion systems. In this chapter we will evaluate technologies that either act externally or use electromagnetic energy to enhance the overall launcher performance.

3.1 Reduction of Required Δv

Referring to the exponential factor determining the payload mass fraction in Tsiolkovski's eq. (1.7), a small decrease of the required Δv to reach orbit will increase payload capability overproportionally. For example, let's assume a standard LO_2/LH_2 engine with an I_{sp} of 450 seconds, and a $\Delta v = 8,000$ m/s to reach LEO orbit. Then the payload mass fraction would be 16.3% of the total launcher mass. If we now reduce the required velocity increase by only 300 m/s, the payload mass fraction would rise to 17.5% which is an increase of about 7%.

This can be achieved by several means:

 (i) Launching from an aircraft with an initial velocity;

 (ii) Providing an initial boost with a chemical/electromagnetic catapult;

 (iii) Launching outside the atmosphere on top of an ultra-high tower.

3.1.1 Aircraft Assisted Launch

Launching from an aircraft produces two advantages: first, an initial velocity and second, if the altitude is high enough, a reduction of the drag losses encountered in the atmosphere. One example is the Pegasus booster rocket from Orbital Sciences launched from the L-1011 aircraft (Fig. 3.1). Typical aircraft performance of a Boeing 747 aircraft is a maximum cruising speed of 0.85 Mach (= 255 m/s) and an altitude of 13 km. Although drag reduction at this altitude is not so important, the initial boost when separating from the aircraft is close to our example at the beginning of this chapter.

Fig. 3.1. L-1011 aircraft with Pegasus booster (Courtesy of NASA)

Fig. 3.2. Space Shuttle Enterprise on top of Boeing 747 (Courtesy of NASA)

The Pegasus rocket (launch mass 22 t) is very small compared to a real SSTO launcher (e.g., Space Shuttle orbiter weights 104 t, the complete launcher 2000 t). Hence launching a big rocket would greatly increase complexity and costs. Also separation would be much more difficult since the launcher must be carried on top of the aircraft and not below as with the small Pegasus (see Fig. 3.2). Supersonic speeds and higher altitudes are also a costly factor. However, for small rockets/payloads, aircraft-assisted launch is an option which is worthwhile studying.

3.1.2 Catapults

The idea of launching payloads into orbit using a giant gun dates back to the Jules Verne story *Voyage from Earth to the Moon* in 1885. Not only classical guns, also gas guns, electromagnetic accelerators or other catapult concepts are in principle capable of bringing payload into orbit – or providing an initial boost velocity to reduce the launcher's Δυ requirement.

Obviously, although sufficient speeds can be obtained, there are some difficulties which have to be considered:

(i) Very high accelerations: Depending on the length of the gun and on the velocity one wants to achieve, the accelerations can be as high as 2,000 up to 100,000 g (Earth gravitational acceleration) for launching directly into orbit. No biological organism can withstand such acceleration loads. Also spacecraft hardware has to be specially protected adding significantly to mass, complexity and costs. Low accelerations appropriate to humans (3 g) reduce the initial boost that can be achieved, or require very long tubes. The relationship between catapult length l, exit velocity v and acceleration a is given by

$$a = \frac{v^2}{2 \cdot l}. \tag{3.1}$$

(ii) Using our example above, for an initial boost velocity of 300 m/s, the required catapult length is 1.5 km for an acceleration of 3 g. A 100 m-long gun that provides a Δυ = 8,000 m/s to launch payload directly into LEO orbit has an acceleration of 32,600 g!

(iii) Heat shields: High velocities through air cause significant drag which requires a heat shield. Such shields can be reusable or ablative; however, they add to the launcher mass therefore reducing the payload capability. Also heat transfer towards the payload may need to be considered.

(iv) Environmental issues: Guns produce a shock wave, and therefore operating personnel and surroundings have to be protected. A remote infrastructure is required which increases the logistic complexity.

Moreover, although a variety of catapults have been built, their size was always geared to propelling of projectiles in the order of kilograms. Upscaling catapults to accommodate projectiles with hundreds of tons adds significant complexity and costs. In general, catapults can be subdivided into thermodynamic (including chemical, pneumatic) and electromagnetic ones. Catapults may not only be used to provide launch assists from Earth. Electromagnetic catapults, for instance, can be used to launch payload from the lunar surface towards Earth since chemical propellants are very expensive on the moon but a nuclear power plant installed to provide the necessary energy is always possible.

Gun Launch

A classical cannon consists of a tube with thick walls and uses a high-energy explosive (usually solid) which is ignited at the closed end of the tube (Fig. 3.3). After ignition, a high-temperature and high-pressure gas expands down the length of the tube accelerating the projectile. Typical gas temperatures are 3,500 K, pressures as high as 3,500 bar. This requires a special protective shell around the spacecraft. Accelerations are within 10,000 and 40,000 g and are therefore not suited for humans. Large, long-range powder guns were used in World War I (Paris Gun) and during the 1960's for the High Altitude Research Program (HARP) in the United States (see Fig. 3.4). The Paris Gun, also known as Big Bertha, could propel a 120 kg shell to an altitude of 40 km at the edge of space. This was the highest altitude attained by a man-made object until the first successful V-2 flight test on October 3, 1942. The HARP cannon holds the record firing a 85 kg projectile up to an altitude of 180 km. The final exit velocity depends on the speed of sound in the driver gas. Due to the relatively high molecular weight of the combustion gases, the upper limit is about 3 km/s of the initial boost velocity.

The constructor of HARP Gerard Bull was contracted by Iraq in the late 1980's to develop the supergun *Project Babylon* with a diameter of 1 m. This gun would have had the capacity to put a 2,000 kg rocket-projectile into a 200 km orbit at a cost of 600 $/kg. Bull was assassinated by the Israelis in 1990, not due to his work on guns but due to consultancy with Iraq about multi-stage missiles, and the supergun was never completed.

Gas Gun Launch

The gas gun circumvents the speed of sound velocity limitation by using a highly compressed low molecular weight gas (e.g., hydrogen or helium) instead of solid explosives. By injecting the gas along the tube it is possible to reduce the peak pressures down to 1,000 bar (Fig. 3.5). Because the high pressure has to be maintained along the whole tube length, the continuous injection becomes a problem if a very long tube is used for moderate acceleration levels. The largest gas gun SHARP (Super High Altitude Research Project) is operated at the

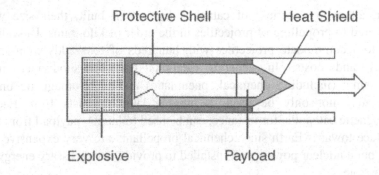

Fig. 3.3. Gun launch

Fig. 3.4. HARP gun (Courtesy of Peter Millman)

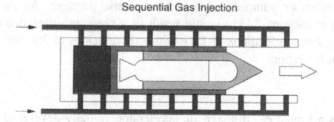

Fig. 3.5. Gas gun launch

Lawrence Livermore National Laboratory in the United States and is capable of accelerating a 5.8 kg projectile to 2.77 km/s. A full scale gun was under consideration with an exit velocity of 7 km/s and a projectile mass of 5,000 kg. The payload would be 1.7 m in diameter and 9 m long. Following burn of the rocket motor aboard the projectile, a net payload of 3,300 kg would be placed into LEO. However, after 1 billion $ of funding, space launch tests of small projectiles at 7 km/s were not forthcoming.

Ram Accelerators

In this concept, the tube is filled with a pre-mixed fuel and oxidizer (e.g., methane and oxygen). The projectile is injected with an initial velocity and compresses the gas. At the end of the projectile, the gas is ignited and combustion starts. This is

Compression Zone

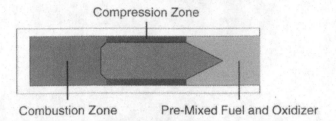

Combustion Zone   Pre-Mixed Fuel and Oxidizer

Fig. 3.6. Ram accelerator

very similar to the operation of a ram jet inside a tube, hence it is called ram accelerator (Fig. 3.6). Because fuel and oxidizer are pre-mixed, problems of mixing at supersonic speeds are avoided. Segments with different mixing ratios are possible to gradually increase the speed of sound in the combustion gas and to obtain a higher exit velocity. Cooling, and the design of the protective shell are adding to the complexity of this system. Experiments have been carried out with projectiles of 4.29 kg, which could be accelerated up to 1.48 km/s.

Pneumatic Catapult

A novel concept employs the atmospheric pressure difference between sealevel and higher altitudes to push the projectile through the tube. In this case, the tube would be placed alongside a mountain. The lower end of the tube would be closed and the upper end vented or pumped down to atmospheric pressure. As an example, a difference in altitude of 2.1 km would result in a pressure difference of 0.25 bar. Studies show that accelerations of 1.5 g with exit velocities of 300 m/s are possible for such a configuration.

Rail Gun

A rail gun is a simple electromagnetic accelerator. Initially developed for weapon applications mainly during the US Space Defence Initiative (SDI) program in the 1980's, they can also be used to provide launcher assists. It consists of two

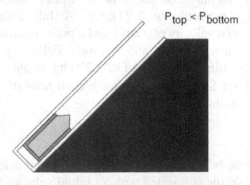

$P_{top} < P_{bottom}$

Fig. 3.7. Pneumatic catapult

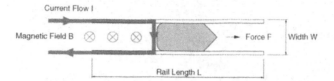

Fig. 3.8. Rail gun principle

conductive rails separated by the distance W with a projectile between them connected to a power supply. The current flow is either closed by the conductive projectile, or by a plasma generated behind the projectile. If a current is flowing along the path shown in Fig. 3.8, a magnetic field $B = \mu_0 \cdot I$ is generated normal to the current flowing plane. This causes magnetic forces on the rails themselves pushing them away from each other and a force on the projectile moving it out of the rail. The force on the projectile is expressed by

$$F = B \cdot I \cdot W. \tag{3.2}$$

The exit velocity of the projectile can be calculated by balancing the kinetic energy and the energy gained by the magnetic force along the rail

$$\frac{m \cdot v^2}{2} = F \cdot L. \tag{3.3}$$

Using the expression for the magnetic field and force, we can isolate the velocity as

$$v = \sqrt{\frac{2FL}{m}} = I \cdot \sqrt{\frac{2LW\mu_0}{m}}. \tag{3.4}$$

Therefore, the velocity is only a function of the current, the projectile's mass and the geometry of the rails. However, as the projectile moves along the rail, a current is back induced causing a decrease of circuit voltage V across the rails. This limits the maximum obtainable velocity to

$$v_{max} = \frac{V}{BL}. \tag{3.5}$$

Efficiencies are quite moderate, between 40% and 70%. The energy is typically stored in high voltage capacitors, which are discharged through the rail-projectile assembly. Top rail guns can presently accelerate a 2 kg payload to a velocity of 4 km/s along 6 m rails separated by 5 cm. Using Eq. (3.4), the required current for such a "small-scale" launch assist catapult is already 6.5 million Ampere! This of course causes serious damage to the rails due to heating as this very large current passes through. Many railguns of this size could only be used for one shot. The high

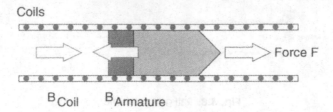

Fig. 3.9. Mass driver (coil gun) principle

current need is due to the fact that the magnetic field generated by the rail is very weak – it is similar to a coil but with only one wind round. This can be circumvented by using large coils, permanent or superconducting magnetics to assist the magnetic field generation, lower the electric current needs and heat production in the rails.

Mass Driver (Coil Gun)

A mass driver, or coil gun, uses stationary electromagnets stacked together to form a kind of tube. Inside is the projectile with a conductive armature at the end which can move inside the tube. The coils are energized sequentially. If current flows through one coil, a magnetic field is formed in the direction of the tube. This induces a current of opposite direction in the conductive armature, which creates a magnetic field opposite to the coil magnetic field. The magnetic repulsion force moves the assembly to the next coil, which is energized and the assembly is again accelerated. The sequence can be computer-controlled, thus accelerating the projectile to ever increasing velocities. Moreover, it is also possible to de-accelerate the assembly by changing current polarity of the coils in the event of malfunction. The force can be increased by attaching permanent magnets to the armature. Efficiencies are very high (> 90%) if superconducting coils are used. Up to now, only small laboratory prototypes exist, a 340 gram projectile could be accelerated to 410 m/s. However, up-scaling of a mass driver should be much more easy than e.g., railguns. Also acceleration level can be very well controlled, enabling human transportation using such launch-assist technology.

Magnetic Levitation – Maglifter

The Maglifter is a combination of magnetic levitation (maglev) technology using superconducting magnets and a mass driver. Superconducting magnetic levitation is a technology already developed for high speed trains (e.g., the Transrapid in Germany) and is thus mature enough to also accelerate very heavy spacecraft. By proper triggering of the magnets, the spacecraft is then accelerated with far less friction than inside a standard mass-driver tube. NASA is currently building a prototype as shown in Fig. 3.10. Such a system is envisioned to provide an initial boost velocity of 300 m/s at moderate

Fig. 3.10. Maglifter model (Courtesy of NASA)

acceleration levels (< 3 g) at low costs since much of the technology has already been developed.

3.1.3 Ultra-high Tower

One way of reducing launcher Δv requirements is to save the several hundred m/s loss due to air drag by launching outside the sensible atmosphere. This idea goes back to science fiction writer *Sir Arthur C. Clarke* (Odyssee 2001; he also "invented" the Geostationary Orbit for telecommunication satellites), who had the idea of ultra-high towers from which rockets could be launched. In order to significantly reduce air drag, such a tower has to be at least 50 km high. In principle, this might be possible by using graphite-epoxy construction materials. More extreme concepts are so-called Skyhooks, an elevator hanging down from GEO to pull payload into orbit without any need of a launcher. Not only the significant length and strength of the material, also heating and control dynamics are still unsolved questions especially with regards to storms in the atmosphere. However, even if such concepts appear unlikely for application on Earth, they might be an intelligent and cost-effective solution on smaller moons in the solar system.

3.2 Advanced Drag Reduction

Air-drag friction can be reduced by a number of methods, such as aerodynamical vehicle shapes and flexible surfaces. All classical principles try to modify the flow in such a way that turbulence is minimized. Obviously, the best reduction would be achieved if airflow towards the vehicle could be reduced. Advanced concepts show that this is indeed possible involving the use of electric charges and generation of plasmas.

3.2.1 Surface-charged Vehicles

A very simple way of generating a vacuum in front of a body is to charge it! Let us suppose a conductive surface is connected to an electrostatic generator. Depending on the polarity, the surface will try to collect electrons from, or implement electrons into, the ambient air molecules (mainly consisting of nitrogen and oxygen). This generates ions with the same polarity as the charged surface, and are therefore repelled by it. Since the air is pushed away, the air pressure close to the surface is lower than the ambient one. The vacuum quality and hence drag reduction can be controlled by the surface potential limited by on-board power availability.

Close to the Earth's surface, air is slightly positive ionized. However, as we move up in altitude, negative air ions are much more dominant (due to solar radiation). To obtain optimal drag reduction, the charge therefore has to be optimized to the flight profile of the launcher. The concept is illustrated for a positively-charged surface in Fig. 3.11. Such a concept was patented already back in 1963 (H. Dudley, *Apparatus for the Promotion and Control of Vehicular Flight*, US Patent 3.095.167, 1963). Dudley reported that if model rockets were charged, an increase in altitude of 500–600% up to 180 meters could be achieved. In addition, the flight profile appeared much more stable. Wind tunnel and flight testing of real launcher structures is necessary to determine the amount of charge (and therefore on-board power) needed to obtain efficient air drag reduction.

3.2.2 Energy Spike

It is well known that a pointed noise in contrast to a blunt nose decreases the drag load towards the vehicle considerably. This is applicable to aircraft at hypersonic speeds up to several Mach. However, a spacecraft needs to reach about Mach 25 for

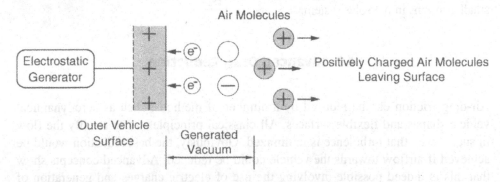

Fig. 3.11. Vacuum generated by positively-charged surfaces

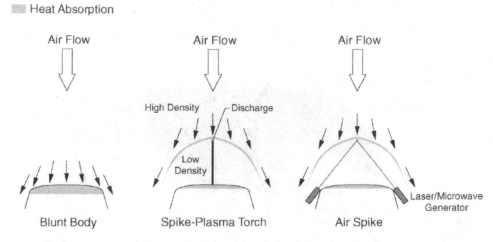

Fig. 3.12. Energy spike for drag reduction

orbit insertion. Since a pointed nose receives all major heat at its tip, where the insulating heat shield is very thin or non-exiting, such high launcher velocities would cause a burn through and destruction of the heat shield. That is why vehicles gaining very high velocities have a blunt nose, distributing the heat load over the surface at the cost of increased drag.

If the heat transfer zone could be moved away from the vehicle, much less heat would have to be absorbed by the blunt nose. One way of doing this is to expel part of the rocket exhaust in front of the vehicle. This technique is actually used for very high speed Russian torpedoes (Shkval) in water. However, for very high velocity flight this seems not to be practible. Another idea is to trigger a very hot plasma discharge in front of the vehicle on a spike (plasma torch) to generate a shock wave that will redirect the air flow and absorb part of the heat load. An illustration is shown in Fig. 3.12. Inside the bow shock, the air has a lower density reducing the drag load to the blunt surface. An even more radical concept was suggested by Leik Myrabo and Yu Razier in that a laser or microwave generator is used to focus energy and create a plasma discharge similar to the plasma torch, but avoiding the physical spike structure (and avoiding the heat load to it). Although the principle is feasible, the amount of power necessary to operate such lasers is much too high at the moment to be practical.

3.3 Magnetohydrodynamic (MHD) Propulsion

In addition to air-breathing propulsion concepts such as Rocket-based combined-cycle (RBCC) engines discussed in chap. 2, electromagnetic energy can be utilized

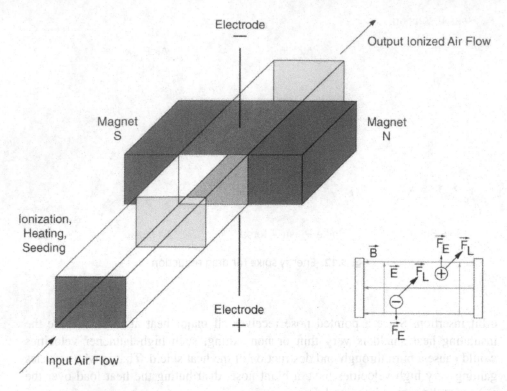

Fig. 3.13. Principle of magnetohydrodynamic (MHD) accelerator/generator

to generate thrust using ambient air as well. This principle is called Magnetohydrodynamic (MHD). Looking at the configuration shown in Fig. 3.13, a current is passed through a conductive fluid or gas (e.g., ionized air). A strong magnetic field is located perpendicular to the current flow. This generates a Lorentz force on each ion which can be expressed by

$$\vec{F}_L = I \cdot (\vec{l} \times \vec{B}), \qquad (3.6)$$

where I is the current, l the length of the channel (in the direction from $-$ to $+$ electrode) and B the magnetic field. Depending on the amount of current passed through the channel and the magnetic field strength, the fluid/gas will be accelerated out of the chamber creating a thrust. Even submarines have been proposed that are silently propelled by MHD, using no moving parts (shown in the famous movie *Hunt for Red October*). To estimate the electrical power requirements, we can substitute the current I in Eq. (3.6) using Ohm's law and get

$$\vec{F}_L = \sqrt{\frac{P}{R}} \cdot (\vec{l} \times \vec{B}). \qquad (3.7)$$

This shows, that the resistance R has to be as low as possible to obtain high thrusts at reasonable power levels. In the case of air-breathing propulsion, this is done by heating up ionized air to create a hot temperature plasma. In addition, alkali metals (e.g., Calium or Cesium) are seeded to increase the ionization fraction and lower required temperatures. A good electrical conductivity of $\sigma = 50-100\,\Omega^{-1}\mathrm{m}^{-1}$ is obtained for ionized air at 2800 K with Cesium seed. For example, let us consider a typical launcher such as the European Ariane 5 rocket. The lift-off thrust is about 6.7 MN. If we want to estimate the electrical power, we need to increase the specific impulse (and therefore thrust assuming a constant mass flow rate) by 10%, we transform Eq. (3.7) into

$$ P = \frac{1}{\sigma} \cdot \left(\frac{F}{l^2 \times B} \right)^2 \qquad (3.8) $$

and obtain $P = 17\,\mathrm{MW}$ for a channel length of 1 m, superconducting magnets providing 20 T and a realistic conductivity of $\sigma = 65\,\Omega^{-1}\mathrm{m}^{-1}$. This amount of power can only be delivered by an on-board nuclear power generator, which is very difficult to integrate into the launcher. Also the magnetic field of 20 T is very optimistic and very difficult to implement.

It is interesting to note that the physical principle of an MHD accelerator can be reversed to generate power from an incoming ionized airflow, reducing its velocity. As we have seen above, considerable power levels can be drawn from such a device, which could be utilized for other propulsion systems or payloads. However, this only works during the period the launcher is inside the atmosphere.

3.4 MHD Energy Bypass Application

Scramjet engines suffer from problems of mixing fuel and air at high Mach number speeds. This can be augmented by using a Magnetohydrodynamic (MHD) generator to extract energy and thus slow down the air flow before entering the combustion chamber. The energy extracted can be reinjected into the beam by using a MHD

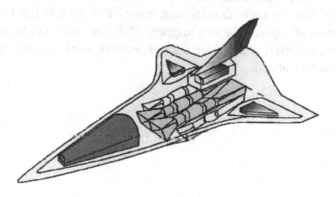

Fig. 3.14. AJAX hypersonic vehicle (Courtesy of Hypersonic System Research Institute)

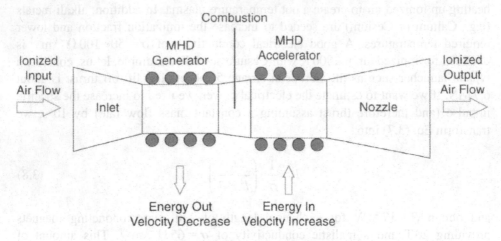

Fig. 3.15. SCRAMJET–MHD energy bypass

accelerator after combustion, which speeds up the airflow. In order to operate a MHD, the incoming air must be ionized. If ionization takes place in front of the aircraft, air drag is also significantly reduced. This concept of combining advanced drag reduction, scramjet and MHD generator/accelerator was suggested by the Russian Hypersonic System Research Institute (NIPGS), the vehicle given the name AJAX (see Figs. 3.14 and 3.15).

Ionization of air is accomplished by using a microwave generator on board the vehicle. Since the MHD generator produces the high energy input needed by the MHD accelerator, the problem of the high energy requirement of MHD propulsion is cleverly bypassed. Superconducting coils would be necessary for the MHD devices in order to operate effectively. Although the generator/accelerator combination introduces some losses due to a certain lack of efficiency (losses are estimated to be 10%), the gain in combustion performance is significant. Calculations for an AJAX vehicle flying at Mach 12 show an energy bypass ratio of 0.28 and increase the specific impulse by 20%. The application of AJAX technology to a SSTO launcher would be very attractive. Numerical studies and MHD test facilities are under development mainly at NASA in the United States. The combination of superconducting magnets, MHD and scramjet technology on a launch vehicle, however, will require a lot of research work in many fields before the concept can be realized.

Chapter 4

Nuclear Propulsion Systems

4.1 Overview

The most powerful energy known to humans is that stored in the nucleus. After fission or fusion of atoms, the end product can have a smaller mass than the initial atoms (fission processes require heavy nucleons such as Uranium, while fusion processes require very light nucleons such as Helium). This mass defect is directly transferred into energy according to Einstein's famous equation $E = mc^2$. Nearly all gained energy is released as heat. While fission or fusion transfer only part of the nuclear binding energy into heat, matter-antimatter annihilation (for example proton-antiproton or even hydrogen-antihydrogen) can release all nuclear energy. Although this would represent the highest energy density achievable, production and storage of antimatter is at a very early development stage. Figure. 4.1 compares the typical energy stored in the chemical reaction between hydrogen and oxygen, fission of Uranium 235, fusion between Deuterium and Tritium, and Proton-Antiproton annihilation processes.

The energy stored in nuclear propellants is an enormous $10^7 - 10^9$ times higher than that of optimally performing chemical propellants. A propulsion system utilizing nuclear energy must be therefore capable of achieving nearly any specific impulse close to the speed of light. On the other hand, nuclear processes use very small quantities of matter. If nuclear reaction products are used only, the resulting mass flows and thrust levels would be very low. In order to achieve high thrust levels, a working fluid/gas (or the complete spacecraft mass itself as we will see in one concept) needs to be coupled with the gained nuclear energy compromising the specific impulse. Also material constraints (melting temperature of heating chamber, ...) need to be considered, which may be relaxed by using magnetic bottles to confine the hot plasma. In principle, all combinations between specific impulse and desired thrust level can be engineered. This offers the possibility for undertaking, e.g. manned solar system exploration in acceptable time frames. In fact, nuclear propulsion is the only choice of mature and developed technology that can offer high specific impulse and high thrust levels simultaneously today. A nuclear rocket was developed near completion during the 1960's in the United States NERVA program. Renewed interest in manned Mars missions may revive efforts in nuclear propulsion despite present environmental and political considerations.

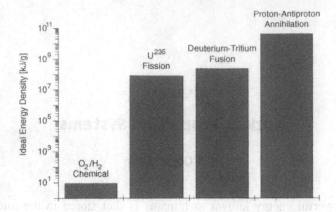

Fig. 4.1. Propellant energy densities for chemical and nuclear propulsion

4.2 Fission Propulsion

Nuclear fission is achieved by bombarding a nuclear core with neutrons, which are not affected either by electrons or protons. The neutron then deforms the core, creating a metastable nucleon. Electrostatic repulsion causes the nucleon to split into two pieces. Since the binding energy of heavy nucleons is much smaller than for lighter ones, the sum of the split product masses needs to be smaller than the initial mass in order to obey the energy balance. This difference in mass is released energy, which can be used for power or propulsion purposes. In addition to the split product, several neutrons are released as they are no longer used by the initial nucleon. These neutrons can then trigger additional nuclear fission leading to a chain reaction. This can be controlled by the speed of the neutrons. The slower the speed, the higher the probability is of additional fission. Hence heavy water or graphite material is used to slow down neutrons by collision processes and control chain reactions inside the reactor.

A typical fission reaction of Uranium 235 (relative atomic mass) is shown in the following equation:

$$n + {}^{235}_{92}U \rightarrow {}^{140}_{55}Cs^* + {}^{94}_{37}Rb^* + 2n + 200\,\text{MeV}.$$

4.2.1 Historical Overview – NERVA/PLUTO Program

The US Atomic Energy Commission (AEC) and NASA joined together in 1960 and founded the Space Nuclear Propulsion Office (SNPO). In 1961, an industrial

Fig. 4.2. NERVA engine mockup (Courtesy of Westinghouse)

contractor team consisting of Aerojet General Corporation and the Astronuclear Laboratory of the Westinghouse Electric Corporation started to workon the Nuclear Engine for Rocket Vehicle Application or NERVA program (see Fig. 4.2). During this program, twelve engines were developed and tested, building upon experience gained from the KIWI reactor and the ROVER program (this was only the reactor and not a rocket) started by Los Alamos National Laboratories already in 1955. Two types of solid core nuclear thermal rocket engines using hydrogen as propellant were investigated, the NRX and Phoebus series (see Table 4.1). These high thrust and high specific impulse engines were directed towards manned exploration such as that of Mars. The US moon landing program of the 1960's could also have benefited from this development. However, the use of nuclear propulsion would have delayed the first human step on the moon and was therefore not considered. Test firings were conducted at the Nuclear Rocket Development Station in Nevada. Initially, rockets were fired into the atmosphere. Due to massive protests against air and ground nuclear bomb testing, the atmospheric test ban treaty was issued in 1963. Although the NERVA nuclear rockets kept the radioactive fission products inside the reactor and not in the exhaust plume, expensive test facilities redesigns had to be implemented in order to filter and store all gases.

Table 4.1. NERVA nuclear rockets

	NRX Series	Phoebus Series
Power	1,500 MW	4,500 MW
Mass Flow Rate $\dot{m}_p$	42 kg/s	129 kg/s
Thrust	333 kN	1,112 kN
Specific Impulse	825 s	820 s
Total Engine Mass	6,800 kg	18,150 kg

Impressive results were obtained and even an engineering model of the NRX series consisting of all peripherals such as turbo pumps, cooled nozzles and valves was tested pointing downwards simulating actual space conditions (see Fig. 4.3). This test occurred in September 1969, was started and stopped 28 times and operated for a total of three hours and 48 minutes with 11 minutes at full power (333 kN!). The next step would have involved flight testing of this engine. However, after spending 2.4 billion US $, the NERVA program was cancelled in 1971 because NASA had no scheduled requirement for its capability.

Another revolutionary concept was designed and tested in the 1960's – a nuclear-powered ramjet engine. After the Sputnik shock, the United States were

Fig. 4.3. NERVA NRX-XE engineering model (Courtesy of Westinghouse)

Fig. 4.4. Tory-IIC nuclear ramjet engine mounted on railroad car in 1964 (Courtesy of LLNL) Credit is given to the University of California, Lawrence Livermore National Laboratory, and the Department of Energy under whose auspices the work was performed

concerned that the Russians could also be ahead in anti-missile technology. To counter that threat, the Pentagon decided to develop a nuclear-powered cruise missile able to fly below enemy defenses. In 1957, the Lawrence Livermoore National Laboratory was picked to develop a nuclear ramjet engine for this weapon given the codename Pluto. A nuclear ramjet uses air as propellent, which is heated by nuclear energy. This gives a basically unlimited range since neither propellant or energy is limited. Of course, such a system would be very attractive also for a rocket-based combined-cycle engine, drastically decreasing propellant needs during launch. The big design challange was to design a very compact reactor core which could fit on a missile withstanding high temperatures, very high speeds and able to work under all weather conditions. In 1961, the engine protype Tory-IIA was fired for a few seconds successfully at a fraction of its power. In 1964, Tory-IIC was mounted on a railroad car (see Fig. 4.4) and fired for five minutes at full power of 51 MW, giving a thrust of 155 kN. Despite these successes, the project was stopped shortly afterwords due to concerns about nuclear contamination along the flight and the projected costs for the missile.

4.2.2 Solid Core

The NERVA program implemented a solid core nuclear reactor as a heat source to accelerate a working fluid (in most cases hydrogen, some concepts involve ammonia) and expel it through a nozzle (see Fig. 4.5). The KIWI fuel element contains Uranium carbide particles as the fissionable element coated by graphite. The flow passages

Fig. 4.5. Solid core nuclear rocket

were covered with Niobium to protect the graphite from the corrosive effects of the hydrogen propellant. Due to material temperature constrains (about 2650 K), the maximum achievable specific impulse is in the order of 900 s. In principle, the rocket and the fuel elements (with fissionable material below the critical mass) can be launched into space separately and assembled (e.g., near the International Space Station) in orbit to reduce risk of nuclear contamination. By replacing the nuclear core (which is more or less independently developed for nuclear submarines, etc.) with a non-nuclear heat source, all engine components can easily be tested without the need of building a dedicated nuclear testing facility. Since the NERVA program, such nuclear rockets are at a final development stage and could be readied for flight applications in a short amount of time. As the thrust to weight ratio is greater than one, a launcher also could be equipped with nuclear thermal rockets. Using our example of a Single-Stage-To-Orbit vehicle (see chap. 1) and a chemical LO_2/LH_2 engine with a payload fraction of 10 – 15%, a SSTO launcher with nuclear propulsion could drastically increase the payload fraction to 45%!

The performance of such a solid core nuclear rocket can be further improved by packing spherical fuel elements together instead of using large single fuel elements. Hydrogen can then flow between the fuel spheres with a much higher surface-to-volume ratio which provides superior heat transfer. The main advantages are higher possible operating temperatures (up to 3000 K) and thus higher specific impulses and a more compact and lighter design. This derivative is called Particle-bed Reactor (PBR) and can increase the thrust-to-weight ratio up to 40:1 compared to the NERVA design of only 5:1. The PBR was pursued initially by the US Department of Defense for interballistic missile interceptor applications (Project Timberwind) in the 1980's, and was taken over by the Air Force Space Nuclear Thermal Propulsion program (SNTP), which lasted from 1991 – 1993. Progress was made mostly on a theoretical level but also a single fuel element was successfully tested.

4.2.3 Liquid Core

Instead of using solid fuel particles, liquid nuclear fuel also could be used in a rotating drum configuration as shown in Fig. 4.6. Here, the working fluid could

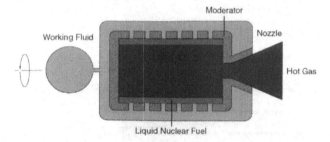

Fig. 4.6. Liquid core nuclear rocket

be heated above the nuclear fuel melting point (but below the moderator melting point), offering higher specific impulses in the order of $1,300-1,500\,$s. The main disadvantage is the loss of fissionable material with the propellant, which results in contamination and increased costs due to resupply of nuclear fuel.

4.2.4 Gas Core

A high specific impulse is estimated for gas core nuclear thermal rockets. In this concept, the nuclear fuel is contained in a high-temperature plasma and radiant energy is transmitted to the propellant. The propellant (e.g., liquid hydrogen) can be used to cool both nozzle and plasma container (e.g., a Quarz cell) in order to achieve specific impulses ranging from $3,000\,$s to $7,000\,$s.

4.2.5 Fission-fragment Propulsion

All previous concepts involved the use of a working fluid heated by nuclear fission with the specific impulse limited by material constraints. Fission fragment propulsion on the other hand uses the energetic fragments produced in the nuclear fission process directly as propellant. Due to the large quantities of energy carried by those products, the propellant velocity and thus the specific impulse is as high as a few percent of the speed of light! In this concept, several graphite fiber discs are covered by an appropriate nuclear fuel and revolve through a reactor core which contains moderators to slow down the neutrons and facilitate a chain reaction. Due to radiation, the nuclear products are heavily ionized and can be directed by strong magnetic fields (see Fig. 4.7). The extremely high specific impulse is achieved at the cost of very low thrust due to the small mass flow rate. Since many discs would have to be mounted together, and heavy and complicated magnets would have to be used, this concept does not seem practical at the present stage.

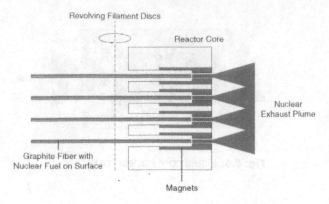

Fig. 4.7. Fission-fragment propulsion

4.2.6 Improvements

LOX-augmented Nuclear Thermal Rocket (LANTR)

Thrust of a nuclear thermal rocket using hydrogen as propellant can be increased by using an oxygen "afterburner" at the beginning of the nozzle (see Fig. 4.8). The oxygen reacts with the hydrogen gas and releases additional chemical energy. Of course, the end product (H_2O) is much heavier than only hydrogen, and thus the velocity (and so the specific impulse) is reduced. As an example, a 67 kN thrust engine with a specific impulse of 940 s can increase the thrust to 184 kN at a reduced specific impulse of 647 s, using an oxidized-to-fuel ratio of 3. As a novel feature, a LANTR provides easy thrust modulation by varying the liquid oxygen flow rate. The O_2 necessary for LANTR mode can also be collected during a mission e.g. from lunar material or by dissociation of CO_2 from the Marsian atmosphere.

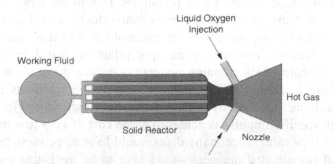

Fig. 4.8. Liquid oxygen augmented nuclear thermal rocket (LANTR)

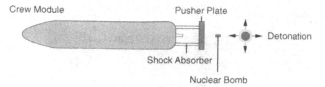

Fig. 4.9. Nuclear pulse rocket

Nuclear Pulse Rocket

Better utilization of the energy yield produced by a nuclear fission reaction is possible by using a nuclear pulse concept. In pulsed operation, much higher exhaust temperatures are possible due to the short interaction time of the propellant with the structure of the vehicle. As an extreme example, a small fission bomb can be dropped behind the rocket (at a certain safety distance) to accelerate it (see Fig. 4.9). In this case, the rocket itself becomes the "working fluid". Of course the blast would have to be transmitted by a shock absorber system to reduce acceleration loads. This idea was the basis of the project ORION which was studied by NASA in the 1960's. The proposed vehicle was 10 m in diameter, 21 meters long and had a weight of 585 tons. Approximately 2000 atomic bombs would have been required for a 250 days' round trip to Mars. Even subscale tests using chemical explosives were performed to prove the concept. However, due to political constrains (no nuclear weapons in space), project ORION was not pursued.

4.2.7 Induction Heating

By using an induction coil around the nozzle, additional electric energy can be used to further heat up the propellant and increase the specific impulse. Waste heat combined with electro-thermal electricity generators (such as Stirling generators or Peltier elements) can be used for this purpose.

4.3 Radioisotope Nuclear Rocket

Radioisotope thermoelectric generators (RTG) are commonly used as an energy source for deep space satellites (e.g., Voyager, Cassini,...). Heat is produced by the nuclear decay of radioactive elements (e.g., Plutonium), which can be directly, converted into electricity. Removing the thermocouple elements, the nuclear fuel can also be used to heat up a propellant such as hydrogen but also water, ammonia or helium (see Fig. 4.10). Typical temperatures are about 1,500 °C to 2000 °C corresponding to a specific impulse of 700–800 s. A small 5 kW nuclear rocket would only weigh 13.6 kg and produce a thrust of 1.5 N

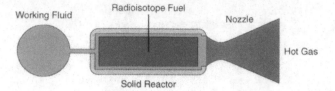

Fig. 4.10. Radioisotope nuclear rocket

applicable to attitude control or planetary missions on a low thrust trajectory. Alternatively to Plutonium, short half-life elements such as Polonium (138 days) can also be used as a nuclear fuel in order to lower nuclear contamination risk. TRW in the United States successfully demonstrated a 65-hour test of a radioisotope nuclear rocket using Polonium as the nuclear fuel and hydrogen as the propellant as early as 1965.

4.4 Fusion Propulsion

If two light nuclear cores are fused together (e.g., hydrogen), the resulting heavier nuclear element has less binding energy than the sum of the two original ones. The energy difference is again mostly released as heat. Comparing the energy densities of heavy element fission and light element fusion, the latter can release nearly an order of magnitude more energy for the same amount of nuclear fuel mass. In fission technology, neutrons are used to make the nuclear core unstable. Fusion is much more complicated because in order to bring the two positively charged nuclear cores close together, the energy of electrostatic repulsion has to be overcome and also maintained. In addition, the cross section for nuclear fusion is very low, which requires a high density of nuclear cores in order to obtain sufficient fusion reactions. A fusion reaction which has a relatively high cross section and requires low initial energy is achieved by Deuterium (hydrogen with 1 neutron) and Tritium (hydrogen with 2 neutrons)

$$^2_1D + {}^3_1T \rightarrow {}^4_2He + n + 17.6 \text{ MeV}.$$

The initial energy required for this reaction is about 10 keV, which equals a temperature of 75 million Kelvin. Hence the nuclear fuel will be a highly ionized plasma. The heating process can be achieved by many different means such as electromagnetic induction, laser or particle bombardment. The main technical difficulty is to confine the plasma so that it does not touch the chamber walls and maintains a high density. Uncontrolled fusion has been achieved by thermonuclear fusion bombs. Controlled fusion along the development of a fusion power plant is still under heavy development (a famous saying since 40 years is that controlled fusion is 10 years away). A world record has been achieved at the Joint European Torus (JET) facility in the United Kingdom where the first major fusion reaction of Deuterium and Tritium was observed in 1991 (2 MW of thermal energy) and in 1997, when already 60% of the initial energy could be drawn from fusion reactions for a time period of 1 minute. Fusion energy at low gain (energy release well below initial energy $< 1\%$) however is already possible with a variety of different designs.

4.4.1 Inertial Confinement Fusion (ICF)

A pellet containing the fusion fuel (e.g., D-T) is placed in the focus of either laser or particle beams where it is heated and compressed (see Fig. 4.11). The compression can be enhanced by placing high energy explosives (e.g. HEDM material) at the outer shell. Fusion reactions inside the pellet create a high temperature plasma, which can be directed by a magnetic nozzle to create thrust. In order to be effective, superconducting magnets are required for the purpose. The pellet's own inertia is sufficient to confine the plasma long enough so that a useful fusion reaction can be sustained. In a variant called Magnetically Insulated Inertial Confinement Fusion (MICF), an additional metallic shell is added to the pellet. During laser/particle bombardment, electrons leave the surface of this shell and a current flows on its outer surface which creates a magnetic field. This enhances the confinement and leads to a higher fusion power gain.

4.4.2 Magnetic Confinement Fusion (MCF)

Contrary to ICF, a magnetic confinement fusion device only confines the fusion plasma by strong electromagnetic Lorentz forces. For instance, the plasma can

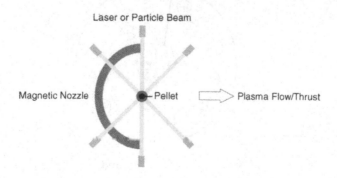

Fig. 4.11. Inertial confinement fusion (ICF)

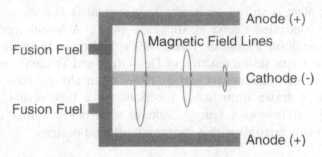

Fig. 4.12. Dense plasma focus thruster

be kept inside a magnetic bottle, heated by high frequency electromagnetic field coils. High energy plasma can then escape through a leak in the magnetic field configuration. Usually, a working fluid is added to decrease the high specific impulse and increase the thrust. Another possibility is to compress, confine and direct the fusion plasma in one magnetic pinch created by a powerful discharge between a cylindrical anode and rod-shaped cathode (see Fig. 4.12). The magnetic field lines concentrate around the cathode increasing the Lorentz force and forming a plasma focus. If the magnetic field is sufficiently dense, fusion reactions will occur and the plasma is expelled through the magnetic gradient from the fuel inlet to the outside (similar to Magneto-Plasmadynamic MPD Electric Propulsion Thrusters). This device is therefore called Dense Plasma Focus Thruster.

Mesh at High Negative Potential

Ground

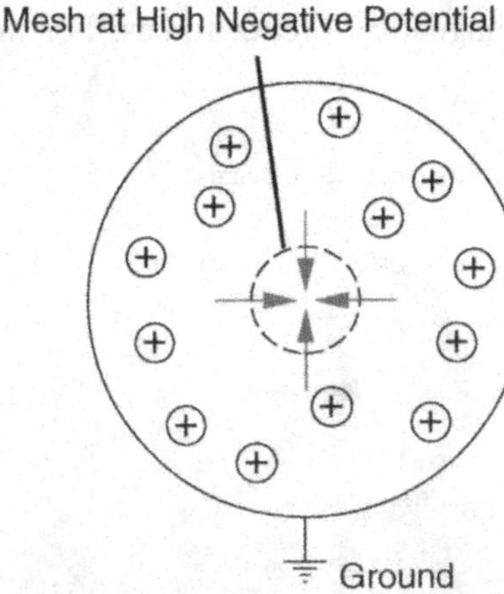

Fig. 4.13. Inertial electrostatic confinement (IEC) fusion

Fig. 4.14. IEC experiment (Courtesy of G. Miley, Fusion Studies Lab, University of Illinois)

4.4.3 Inertial Electrostatic Confinement Fusion (IEC)

Rather than magnetic fields, electrostatic fields can also confine a fusion plasma. This can be accomplished by applying a high electric field between a mesh-shaped electrode and an outer shell (which can also be made out of a mesh) as shown in Fig. 4.13. Positive fusion fuel ions (e.g., D^+, T^+, He_3^+ are injected to the grounded outer shell and accelerated towards the mesh electrode (typically at up to $-100\,kV$ with respect to the outer shell). Inside the mesh, the ions gain enough energy to perform fusion reactions. The plasma can escape from a hole in the mesh electrode to create thrust (see Fig. 4.14). Developed at the University of Illinois in the United States, small-scale IEC fusion devices are already commercially available as portable neutron sources. Upscaling of these devices for propulsion purposes is presently being investigated.

4.5 Antimatter Propulsion

Anti-matter is similar to normal matter but with opposite charge and quantum numbers. For example, an anti-proton has a negative charge and a magnetic moment opposite to the "normal" proton one. If matter and anti-matter collide, a high amount of energy is released in the form of radiation and/or more matter and anti-matter particle pairs. The energy density of this reaction is the highest in nature known up to now (two orders of magnitude higher than ideal fusion reactions). Anti-matter is produced as a by-product in particle accelerators such as CERN in Geneva. If particles collide at relativistic velocities with a target, the energy resulting from slowing down the particle is released in a number of other particles including anti-matter. Obviously, if this anti-matter particle then collides with normal matter, it is instantly destroyed. This is one of the major difficulties in producing anti-matter. Moreover, once anti-matter is isolated from normal matter requiring ultra-high vacuum conditions, it needs to be stored, which is the next technological problem. Even more so if this anti-matter needs to be stored at a density which makes it attractive for further utilization such as propulsion. Trapping devices utilize electromagnetic fields in vacuum chambers cooled down to a few Kelvins by liquid helium (cold anti-matter reduces thermal movement and hence the possibility of colliding with normal matter). Pennsylvania State University in the United States recently developed a portable anti-matter storage device called Penning Trap. It has the capacity to store 10^{10} anti-protons (about 10 femto-grams!). An actual picture of this device is shown in Fig. 4.15.

Not all anti-matter is interesting for direct propulsion applications. For instance the collision of an electron with a positron (anti-matter electron) produces pure

Fig. 4.15. Portable penning trap (Courtesy of Gerald A. Smith, Prof. Emeritus of Physics, Pennsylvania State University)

electromagnetic radiations (photons) in all directions which can hardly be directed:

$$e^- + e^+ \rightarrow 2\gamma + 511\,\text{MeV}.$$

On the other hand, proton–antiproton collisions also produce charged particle pairs which can be directed e.g. by electromagnetic means and directly used for propulsion:

$$p^+ + p^- \rightarrow 3.2\,\pi^+, \quad \pi^- + 1.6\,\pi^0$$

Present costs for one anti-proton is about 10 cents (!) and the yearly capacity of CERN is about 10^{12} anti-protons equal to a nanogram. Improvements could enhance this production rate by a factor of 100. This amount of anti-protons would be just enough to catalyze fission and fusion processes similar to HEDM which enhance chemical propulsion performance. However, propulsion devices which rely only on matter-antimatter annihilation require at least 10^{20} anti-protons, which is far out of reach at the moment. Considering the costs of anti-protons, even anti-matter catalyzed fission/fusion processes are economically not attractive. Nevertheless, the unmatched energy density will make this choice attractive once technological advancements drastically lower costs and increase production rate.

4.5.1 Anti-proton Catalyzed Fission/Fusion Propulsion

This concept uses anti-protons compatible to present production rates and is under development at the Pennsylvania State University in the United States. A pellet consisting of Uranium or Plutonium as fission nuclear fuel and Deuterium–Tritium as fusion nuclear fuel is compressed by either laser or particle beams. The ratio of Uranium to D-T is about 1:9, the total weight of the pellet approximately 1 gram. At peak compression, an anti-proton beam collides with the pellet triggering the Uranium to fission. The energy release during this process then

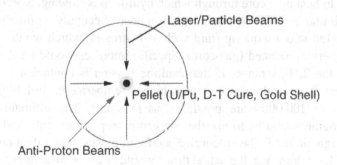

Fig. 4.16. Anti-proton catalyzed fission/fusion propulsion

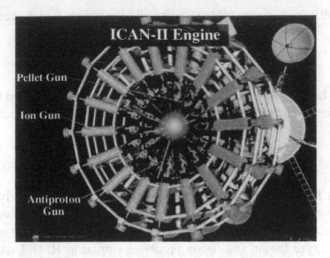

Fig. 4.17. ICAN anti-proton catalyzed propulsion spacecraft (Courtesy of Gerald A. Smith, Prof. Emeritus of Physics, Pennsylvania State University)

drives the fusion of the D-T fuel. This process (fission triggers fusion) is similar to thermonuclear fusion bombs. Comparing this technique to traditional fission/fusion processes, an anti-proton catalyzed fission produces about 6 times more neutrons, greatly speeding up the chain-reaction and hence giving more time to the fusion process, releasing more energy. A concept similar to the nuclear pulse rocket ORION was proposed under the name ICAN (Inertial Confinement Anti-proton Catalyzed Micro Fission/Fusion Nuclear Propulsion). An artist's concept of such an engine is shown in Fig. 4.17.

4.5.2 Direct Antimatter Propulsion System

The possibilities are similar to the thermal nuclear rocket discussed in the nuclear fission chapter. A solid-core antimatter engine would use matter-antimatter annihilation to heat up a core through which hydrogen is flowing. Specific impulses would be similar to NERVA engine performance. Secondly, antimatter could be directly injected into a working fluid such as hydrogen, which would then expand due to the energy produced (gas-core). Specific impulses would be higher and are expected in the 2,500 s range. If the resulting plasma is contained in a magnetic bottle to heat it up and release it at a higher temperature, still higher specific impulses up to 100,000 s are possible (plasma-core). The ultimate performing antimatter engine would be to use the charged pion particles only and direct them by electromagnetic fields (beam-core). Specific impulses would then come close to the speed of light, however, the actual thrust would become close to zero.

Chapter 5

Electric Propulsion Systems

As we have seen in previous chapters, chemical propulsion systems are limited in performance by the energy stored in their propellants. This limitation can be overcome by using electric power available on the spacecraft (e.g., from solar arrays, nuclear reactors or beamed energy sources such as laser light or microwave energy transmitted to the spacecraft) and coupling it with a working fluid or gas. Specific impulses up to 10,000 seconds can be achieved, reducing propellant consumption significantly. In fact, mass savings are so dramatic that electric propulsion thrusters are one of today's enabling technologies in spacecraft technology. They enable much smaller satellites and therefore less expensive launch vehicles, bigger telecommunication satellites and new types of interplanetary missions.

The concept of electric propulsion was already mentioned in the early beginnings by all three space pioneers, namely Robert Goddard in 1906, Konstantin Tsiolkovski in 1911 and further described in Hermann Oberth's famous book "The Rocket to the Planets" ("Die Rakete zu den Planetenräumen") in 1923. Back in 1929, the world's first electric thruster was demonstrated by Vladimir Glushko at the Gas Dynamics Laboratory in Leningrad (now St. Petersburg), Russia. A picture of his simple thruster using electric discharges to vaporize and accelerate the propellant is shown in Fig. 5.1 (as we will see later, this type of thruster is called arcjet).

It was long believed that no spacecraft would ever have enough power available to really use such "advanced propulsion" engines. Even in the 1940's, several publications pointed out that only big and massive nuclear power plants in space would make possible the utilization of electric propulsion. As developments in all space related fields rapidly evolved in the 1950's and 1960's, this point of view changed rapidly. Ernst Stuhlinger, a colleague of Wernher von Braun, wrote one of the major books on this subject (Ion Propulsion for Spaceflight, 1964) stimulating a great deal of research. Solar arrays and compact lightweight nuclear reactors were introduced in satellite designs and the first electric thruster was successfully demonstrated on SERT-I (Space Electric Rocket Test) by the United States in 1964. Russia followed closely by using a similar thruster on an interplanetary probe (Zond 2), and took the lead by using it on many generations of telecommunication satellites. Only recently, interest has again being shown due to the fact that much more power is available on modern satellites and the huge potential mass savings

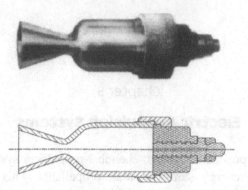

Fig. 5.1. World's first electric thruster built by V. Glushko

associated with electric propulsion. All major space powers including the United States, Russia, Europe and Japan introduce electric propulsion systems in many different satellite platforms; for interplanetary, scientific, microsatellites and most notably telecommunication.

According to their type of application, electric propulsion systems are classified into three basic categories:

(i) *Electrothermal:* Electric power is used to heat up the propellant and expel it through a nozzle. The design is similar to nuclear thermal or radioisotope rockets. Such type of thrusters are commercial off-the-shelf products, and offer moderate increases in specific impulse compared to chemical propulsion systems.

(ii) *Electrostatic:* The propellant is ionized by different means and accelerated by applying a high potential. Very high specific impulses can be achieved at the cost of low thrust levels (μN-N). In addition, superior control of the generated thrust enables new type of space missions which require very precise satellite attitude control.

(iii) *Electromagnetic:* The propellant is ionized by different means and accelerated by an electromagnetic field (magnetic and/or electric forces). This type of thruster combines very high specific impulses and the possibility of processing very high power levels. Such electric propulsion engines can produce much higher thrust levels (N-kN) than electrostatic thrusters.

The ability to transform electric power P into thrust F is expressed by the efficiency η

$$P\eta = \frac{dW_{Kin}}{dt} = \frac{\dot{m}v^2}{2} = \frac{F^2}{2\dot{m}}$$

(5.1)

or

$$\eta = \frac{F^2}{2\dot{m}P}.$$ (5.2)

This is a typical parameter for all electric propulsion systems valid if the initial pressure with which the propellant is fed into the thruster is very small compared to the thrust. In case of very small electric powers, this is not always the case and the efficiency has to be rewritten as

$$\eta = \frac{F^2}{2\dot{m}P + F^2_{\text{Initial}}}.$$ (5.3)

An overview of the performance parameters for the various electric propulsion systems is shown in Table 5.1. Although the thrust levels are very small (mostly sub Newtons), the specific impulses are very interesting. Electric propulsion systems cannot be used for launcher applications, but they are very attractive for spacecraft propulsion systems used for stationkeeping (drag compensation in orbit) and attitude control. Higher thrust engines are even applied as primary propulsion systems for interplanetary spacecraft (e.g., to a comet or the Moon, see low-thrust trajectories in chap. 1) and they can be used very effectively to de-orbit satellites at the end of their lifetime. Typical chemical propellant mass fractions for spacecraft propulsion systems are 20% for LEO, 55% for GEO and up to 85% for interplanetary spacecraft. With a typical specific impulse of "only" 2.000 s (e.g., Ion thruster, Hall thruster) compared to a hydrazine monopropellant engine with 200 s, mass savings are as dramatic as 90% which nearly halfs the size of an interplanetary spacecraft. Depending on the complexity of the electric propulsion thruster, an usually heavy power processing unit (PPU) has to be added to the mass budget. Also the electrical efficiencies of the PPU, which can vary substantially from peak to nominal operating conditions, is an important parameter for a good trade-off between chemical and electric propulsion systems.

We shall note that the propellant does not necessarily need to be on board the spacecraft (as in all concepts discussed in this chapter). One can also envision collecting the neutral gas atmosphere close to a planet (such as in LEO orbit or in a low Marsian orbit).

5.1 Electrothermal

Propellant is heated by electric means and expelled through a nozzle. Since the physical description of this principle is similar to those derived for chemical propulsion systems (only that chemical heating is replaced by electric heating), the exhaust velocity can be calculated according to Eq. (2.5).

Table 5.1. Electric propulsion performance overview

		Propellant	Power	Specific impulse	Efficiency	Thrust
Electrothermal	Resistojet	Hydrazin, Ammonia	0.5–1.5 kW	300 s	80%	0.1–0.5 N
	Arcjet	Hydrazin, Hydrogen	0.3–100 kW	500–2,000 s	35%	0.2–2 N
Electrostatic	Ion	Xenon	0.5–2.5 kW	3,000 s	60–80%	10–200 mN
	Hall	Xenon	1.5–5 kW	1,500–2,000 s	50%	80–200 mN
	FEEP	Indium, Cesium	10–150 W	8,000–12,000 s	30–90%	0.001–1 mN
	Colloid	Glycerol + Additives	5–50 W	500–1,500 s	60–90%	0.001–1 mN
	Laser Accelerated	Any kind	1 MW	10^7 s	?	0–100 mN
Electromagnetic	Pulsed Plasma	Teflon®	1–200 W	1,000 s	5%	1–100 mN
	Magnetoplasmadynamic	Ammonia, Hydrogen, Lithium	1–4,000 kW	2,000–5,000 s	25%	1–200 N
	Variable I_{sp} Plasma Rocket	Hydrogen	1–10 MW	3,000–30,000 s	<60%	1–2 kN

5.1.1 Resistojet

In a resistojet, propellant flows through a heat exchanger connected to an electric resistive heater element (see Fig. 5.2). The heated fuel is then expelled through a nozzle. This very simple and low complex propulsion system can be operated with a variety of propellants such as gases (hydrogen, CO_2, nitrogen,...) or liquids (hydrazine, ammonia,...) and even waste water available on space stations or manned spacecraft. The maximum temperature is limited by the melting temperature of the heat exchanger material such as Tungsten. Typical operating temperatures are in the order of up to 3000 K. This gives a specific impulse of 300 s for hydrazine. Although this is below optimal performing chemical propulsion systems ($H_2 - O_2$ with I_{sp} of 450 s), it is well above secondary propulsion systems like monopropellant or cold gas thrusters (I_{sp} of around 80 s as we will see in a later chapter).

With up to a Newton of thrust, the power consumption of a resistojet is only around a Kilowatt at efficiencies of 80%. This can be even improved by using regenerative cooling instead of radiation cooling to pre-heat the propellant. Resistojets are well developed thrusters (as an example see commercial Resistojets from General

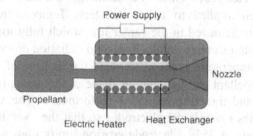

Fig. 5.2. Resistojet concept

Fig. 5.3. MR-501B electrothermal hydrazine thruster (EHT) Resistojet (Courtesy of General Dynamics-OTS, Space Propulsion Systems)

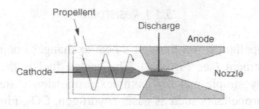

Fig. 5.4. Arcjet concept

Dynamics in Fig. 5.3) and have been used extensively e.g. on the Iridium satellites (first satellite-based mobile phone constellation in the late 1990's).

5.1.2 Arcjet

In order to achieve higher specific impulses, an arc discharge is used to heat the propellant. The propellant is swirled into a rod-shaped cathode and laminar column anode configuration (see Fig. 5.4). The discharge is either generated by applying a low DC voltage (100 Volt) and high current (hundreds of Ampère) or by a high-frequency high voltage field. Also microwave and AC discharges (with the problem of optimizing the electrodes since the discharge oscillates between anode and cathode) have been applied to arcjet thrusters. Temperatures in the order of 30,000–50,000 K are achieved in the center line which fully ionizes the propellant. The heat is transmitted through conductivity and radiation down to the nozzle walls causing a strong temperature gradient to avoid the melting of the nozzle material. Swirling of the propellant and the column shape of the anode increases the time of expose to the arc and therefore improves performance. The very high local heat causes thermal losses towards the electrodes so that the overall efficiency is much lower than resistojets at 35%. Electrode erosion limits lifetime to a typical 1,500 hours of operation.

A variety of propellants can be used depending on specific heat, thermal conductivity and corrosiveness of the thruster materials. Hydrazine (see Fig. 5.5), Hydrogen (highest specific impulse due to lowest molecular weight) or Ammonia are typical choices. Arcjets are well developed and already in operation on a variety of satellites (e.g., on Lookheed Martin Series 7000 Comsat). High power arcjets (up to 100 kW) are presently under investigation in the United States for high Δv missions such as a manned Mars spacecraft.

5.1.3 Solar/Laser/Microwave Thermal Propulsion

A subject that ranges between electric propulsion and beamed energy propulsion is solar/laser/microwave thermal propulsion, which shall be mentioned here. Of course, the propellant can also be heated with an external source of energy such

Fig. 5.5. MR-509 1.6 kW hydrazine arcjet firing (Courtesy of General Dynamics-OTS, Space Propulsion Systems)

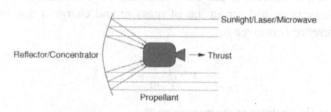

Fig. 5.6. Solar/Laser/Microwave thermal propulsion

Fig. 5.7. Solar thermal upperstage (Courtesy of NASA)

as sunlight or laser/microwave beamed energy (e.g., from an Earth or space-based laser infrastructure). The energy is then concentrated on a heat exchanger or directly on the propellant, which is then heated up and expelled through a conventional nozzle (see Fig. 5.6). Specific impulses of 800–1.200 s and thrusts of several hundred mN are possible using sunlight and hydrogen as propellant. The reflectors can be also made out of inflatable structures as shown in an artist's concept in Fig. 5.7. Laser thermal propulsion offers higher specific impulses but requires a very high pointing accuracy and thus limits its application to very near Earth orbits. Concepts presently under study suggest that using solar thermal propulsion, a telecommunication satellite can be transported from LEO to GEO in about 20 days using very little propellant. This could lower launch costs significantly since much smaller LEO rockets could be used. Low mass inflatable reflectors are also being investigated which would make such an "upper stage" very attractive and feasible.

5.2 Electrostatic

If the propellant is ionized, it can be accelerated very effectively by electrostatic fields. The velocity gained for an ion of mass m and charge q due to the electric potential difference U is given by

$$v = \sqrt{\frac{2qU}{m}}. \tag{5.4}$$

Since the mass flow relates to the current as

$$\dot{m} = I \cdot \frac{m}{q}, \tag{5.5}$$

the force generated by the current I due to the potential difference U can be expressed as

$$F = I \cdot \sqrt{\frac{2mU}{q}}. \tag{5.6}$$

Note that the propellant velocity scales as $(q/m)^{1/2}$ and the force as $(m/q)^{1/2}$ which is important when choosing the propellant. For selecting a very high specific impulse, a multi-ionized, light ion would be ideal. However, since a thruster shall most importantly produce thrust, the propellant of choice is heavy and only singly ionized.

5.2.1 Ion Thruster

A prominent example of electrostatic propulsion systems is the ion thruster. The propellant is either ionized by electron bombardment or through a high-frequency electron excitement, and accelerated using high electrostatic potentials (usually around 1,000 V). Specific impulses are rather high (above 2,500 s), thrust levels between 20 and 200 mN have been demonstrated. Moreover, ion thrusters have already been flown on a number of missions, most prominently on the first interplanetary electric propulsion spacecraft, NASA's Deep Space 1. The biggest disadvantage is the rather high power consumption (several kW), which limits its application. In the beginning, Mercury and Cesium were selected as propellant candidates because they are heavy metals, have a low first ionization energy and a high second one. However, contamination problems with metal propellants and toxity shifted the emphasis towards the use of heavy inert gases such as Xenon or Argon.

Electron-bombardment (Kaufman) Thruster

The electron-bombardment thruster was invented by Prof. Harold Kaufman in the United States. Electrons are emitted into the main discharge chamber and accelerated towards an anode to potentials of typically 1,000 Volts (see Fig. 5.8). The chamber is filled with the propellant gas, usually Xenon, ionized by the energetic electrons. To increase ionization efficiency, a magnetic field is applied to facilitate a gyro-movement of the electrons and thus a higher chance of ionizing a neutral atom along its way to the anode. The ions are slowly pushed through an extractor grid by a small voltage drop between the grid and the anode (typically a few tens of Volt). Then, the ions are accelerated by another grid at negative or ground potential, producing an ion beam with a half-angle divergence of about 10 degrees. Since only positive ions leave the thruster (the electrons inside the main chamber are repelled by the positive extractor grid), and the spacecraft is not grounded in space, the spacecraft potential would continuously become negative. Therefore, electrons have to be injected externally into the ion beam to keep the

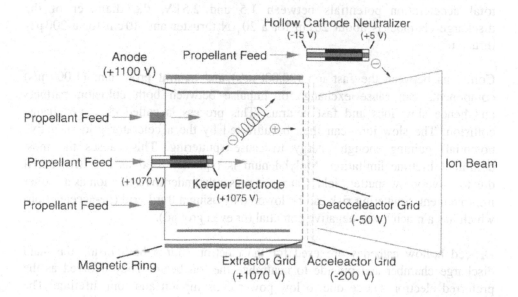

Fig. 5.8. Electron-bombardment (Kaufman) thruster

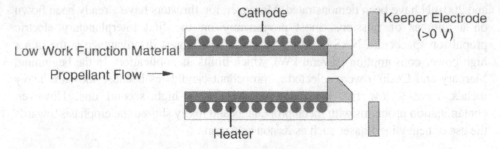

Fig. 5.9. Hollow cathode concept

spacecraft potential constant. Moreover, only positive ions would create a strong positive potential in front of the thruster, which would slow down the ions accelerated from the chamber. If it is high enough, it can even stop the ion current and the thrust would be zero (space charge limitation). The electrons help to create a quasi-neutral plasma that avoid these problems.

Space charge is also a limiting factor for the grid size. The maximum current I that can be extracted from an area A by a potential difference U at a distance d is given by the Child-Langmuir-Schottky law as

$$I = \frac{4}{9} \varepsilon_0 \sqrt{\frac{2q}{m}} \frac{U^{3/2}}{d^2} A. \tag{5.7}$$

A high current (and thrust) therefore requires a high potential difference at a small distance, which can cause sparks. The limit here is usually 0.5 mm. With total acceleration potentials between 1.5 and 2.5 kV, the diameter of the discharge chamber is about 20 cm for a 20 μN thruster and 40 cm for a 200 μN thruster.

Collisions between the fast ion (30,000 m/s) and neutral propellant (1,000 m/s) components can cause exchange of impulse between both collision partners and hence slow ions and fast neutrals. This process is called charge-exchange collision. The slow ions can then be attracted by the accelerator grid (negative potential) gaining enough energy to cause sputtering. This causes the most dominant lifetime limitation. Molybdenum is typically used as a grid material due to its very low sputter yield, Carbon is presently under investigation as a further improvement. Sputtering risk can be lowered by using a third grid (de-acceleration) which has a much lower negative potential (or even ground).

Orficed hollow cathodes are used for the electron source both inside the main discharge chamber and outside to neutralize the ion beam. They evolved as the preferred electron source due to low power consumption and long lifetime. The cathode consists of a low work function material (such as Barium oxide), which is heated to cause thermionic emission of electrons. For its operation, a small part of the propellant gas needs to be flowing through the cathode as well. The electrons hit the propellant gas and ionize it, creating an additional electron. Moreover, the ions are attracted to the low work function insert and cause additional heating. The electrons are extracted by a keeper electrode with a potential difference of a few Volts (typically around 15 V). Once they are accelerated, the neutral propellant, which was not ionized inside the cathode and also flows out of the orfice, is ionized, creating a low energy plasma in front of the cathode. This plasma couples very well with the ion beam and the exact amount of electrons is extracted for neutralization. Such hollow cathodes are also used for spacecraft neutralization in general since they can provide both electrons and ions depending on the spacecraft potential, which neutralize the spacecraft surface. They are then called plasma contactors.

Kaufman ion thrusters were first built at the NASA Lewis Research Center (now Glenn Research Center) in the 1960's. Even 200 kW ion thrusters with a 1.5 m diameter were built and tested (see Fig. 5.10). Electron bombardment ion thrusters were flown on the Spacecraft Electric Rocket Test (SERT) I and II mission in 1964 and 1970 as well as on the Advanced Technology Satellite ATS-6 in 1974. At that time, Mercury and Cesium were used as propellants. Both SERT missions performed very well, SERT-II was operated for as long as 11 years, and the two Mercury ion thrusters accumulated 5,792 hours of full-power thrusting. However, interest in this technology declined until much higher power was available on satellites. NASA used the NSTAR Kaufman thruster (see Fig. 5.11) using Xenon as propellant on the first interplanetary spacecraft that used electric engines as primary in-orbit propulsion system. Deep Space 1 was

Fig. 5.10. 200 kW Mercury ion thruster (1.5 m diameter) (Courtesy of NASA)

Fig. 5.11. NASA NSTAR ion thruster (2.3 kW, 92 mN) (Courtesy of NASA)

launched as recently as 1998. Hughes Space and Communications introduced the first Kaufman thruster in their HP 601 HP satellite bus (XIPS – Xenon Ion Propulsion System thruster), and the first commercial telecommunication satellite PAS-5 was launched in 1997. In Europe, DERA (now Quinetiq) in the United Kingdom is developing electron bombardment thrusters (UK-XX series), one is being used on the ESA ARTEMIS telecommunication satellite. They are also planned to be used for drag compensation on the ESA GOCE Earth observation satellite for launch in 2006. Presently, there is a shift towards the use of hall thrusters instead of ion thrusters on commercial satellites due to their lower power consumption.

Radiofrequency Thruster

An alternative to the electron-bombardment thruster is the radiofrequency (RF) thruster. Here, the propellant gas is ionized by applying a MHz frequency excitation to free electrons that ionize the neutral propellant (see Fig. 5.12). This concept was invented by Prof. Loeb of Giessen University in Germany. The extractor grid (+ 1.5 kV) redirects the ions into the discharge chamber, the accelerator grid (− 1.5 kV) and deaccelerator grid (around ground potential) are similar to the electron bombardment thruster design. The RF thruster offers design advantages (no hollow cathode in the main discharge chamber,...), the RF plasma process makes the RF thruster slightly less efficient than the Kaufman engines. Astrium in Germany is marketing the RIT-XX thruster series (RF Ion Thruster), the RIT-10 generating 10 μN of thrust was flight-qualified on the EURECA mission in 1993. The RIT thruster is also used on the ARTEMIS satellite (together with the UK thruster).

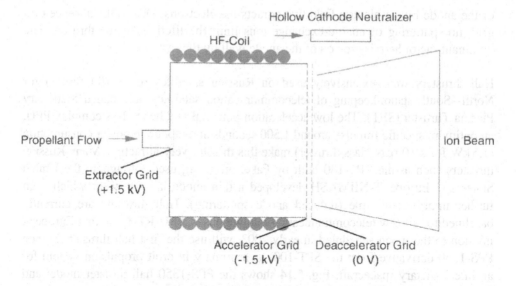

Fig. 5.12. Radio frequency ion thruster

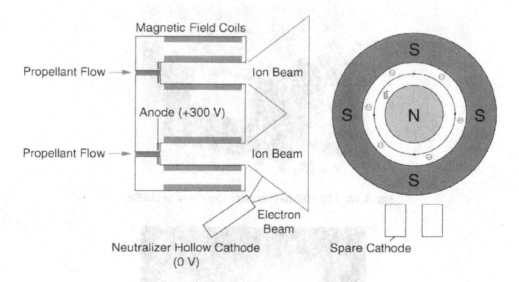

Fig. 5.13. Hall thruster

5.2.2 Hall Thruster

A hall thruster is a gridless electrostatic propulsion system (see Fig. 5.13). Electrons from an external hollow cathode enter a ring-shaped anode with a potential difference of about 300 Volts. Magnetic field coils around the anode ring cause the electrons to spiral inside the ring producing ions. The ions are then accelerated out of the anode by the electric field that attracts the electrons. Due to the absence of a grid, no sputtering of charge-exchange ions limit the lifetime of the thruster. The dominant factor here is erosion of the anode wall by the fast beam ions.

Hall thrusters were extensively used on Russian satellites (over 100 flights) for North–South stationkeeping of telecommunication satellites, and called Stationary Plasma Thruster (SPT). The low acceleration potential and hence less complex PPU, an optimum specific impulse around 1,500 seconds and moderate power consumption (1,5 kW for a 100 μN class thruster) make this thruster very attractive. Many Russian thrusters such as the SPT-100 built by Fakel are being used throughout the United States and Europe. TsNIIMASH developed a thin anode layer thruster which even further improves lifetime (reduced anode sputtering). Hall thrusters are currently baselined for all new telecommunication satellite buses. SMART-1, the first European mission to the moon scheduled in early 2003, will use the first hall thruster (French PPS-1350 derivative from the SPT-100) as a primary in-orbit propulsion system for an interplanetary spacecraft. Fig. 5.14 shows the PPS-1350 hall thruster model and Fig. 5.15 the plume of a Hall thruster in operation.

Fig. 5.14. PPS-1350 hall thruster (Courtesy of SNECMA)

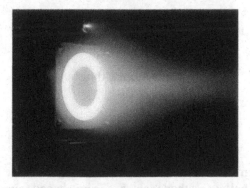

Fig. 5.15. General dynamics BPT-4000 hall thruster firing (Courtesy of General Dynamics-OTS, Space Propulsion Systems)

5.2.3 Field Emission Thruster

A field emission thruster, also called Field Emission Electric Propulsion (FEEP) thruster, uses an electric field to extract and accelerate atomic ions directly from the surface of a metal exposed to vacuum by applying suitable voltages to a closely-spaced electrode configuration. For propulsion applications, the most common source is a metallic liquid. The thrust level ranges from micro- to milli-Newton at specific impulses of as high as 12,000 seconds, requiring total acceleration voltages of up to 10 kV and above. The power-to-thrust ratio is rather high at about 60 – 75 W/mN.

When the free surface of the liquid metal is exposed to a high electric field it is distorted into conical or series of conical protrusions (Taylor cone) in which the radius of curvature at the apex becomes smaller as the field is increased. At a threshold value of 10^9 V/m, atoms on the surface of the tip are ionized and accelerated by the same field that created them, producing a thrust. Due to this direct conversion of a liquid metal into an ion beam, the process operates at high power efficiency. Expelled ions are replenished by the hydrodynamical flow of the liquid metal. A separate neutralizer is required to maintain charge neutrality.

A propellant for such a type of thruster has to have a high atomic mass to achieve sufficient thrust levels, good wetting capabilities to maintain propellant flow to the ion emission site, a low melting point and ionization energy to reduce additional power demands to liquify and ionize the propellant. As a result, the two most common propellants are Cesium and Indium; alternative propellants are Rubidium and Gallium. Other properties such as chemical reactions with ambient plasma or spacecraft components and vapor pressures influencing contamination have to be taken into account, based on the mission scenario.

Two different designs have emerged and are currently under development in Europe (see Fig. 5.16): one based on Cesium or Rubidium as propellant in a slit emitter configuration developed by Centrospazio in Italy and one based on an Indium propellant using a needle or capillary type emitter developed by ARC Seibersdorf research in Austria (see Figs. 5.17 and 5.18). Different reservoir sizes have been developed ranging from 220 mg to 15 g as shown in Fig. 5.19. In the slit configuration, multiple Taylor cone emission sites are automatically built up and are therefore capable of producing high thrusts up to mN. One needle or capillary type can generate a maximum thrust of 100 μN and needs to be clustered for higher thrust requirements (see Fig. 5.20).

Operating down to micro Newton thrust levels with very high thrust controllability enables ultra-high precision pointing capabilities especially for scientific drag-free and interferometry spacecraft (e.g. to detect extraterrestrial planets or gravitational waves). It is also possible to operate a field emission thruster in both continuous and

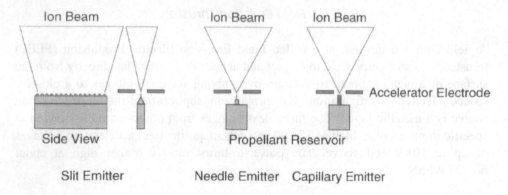

Fig. 5.16. Field emission thrusters

Fig. 5.17. Complete In-FEEP thruster with neutralizer and open electronics (Courtesy of ARC Seibersdorf Research)

Fig. 5.18. Three In-FEEP thrusters firing in vacuum chamber (Courtesy of ARC Seibersdorf Research)

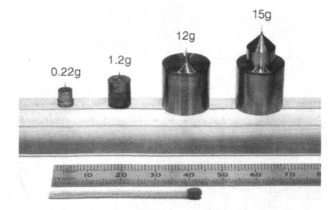

Fig. 5.19. Several Indium ion emitters with reservoirs Ranging from 220 mg to 15 g (Courtesy of ARC Seibersdorf Research)

Fig. 5.20. In-FEEP multiemitter for thrusts up to 1 mN (Courtesy of ARC Seibersdorf Research)

pulsed mode, allowing very small minimum impulse bits below 10^{-8} Ns. Moreover the high specific impulse and compact design (the propellant is stored in solid state) is attractive for attitude control and orbit maintenance of small satellites offering significant mass savings. The high specific power is not a key factor here due to the low thrust level requirements.

Very high accelerator potential needs are certainly one of the most critical elements for spacecraft integration issues. Arcs can cause damage to power supplies and other components. The resulting high mass of the power supplies can easily be one order of magnitude higher than the ion emitter and neutralizer mass. Also electrical efficiencies for such high voltage power supplies are only at about 70%. Certain trade-offs between safety, redundancy and total weight have to be made during the design phase. Another critical element is the use of metal propellants, giving rise to

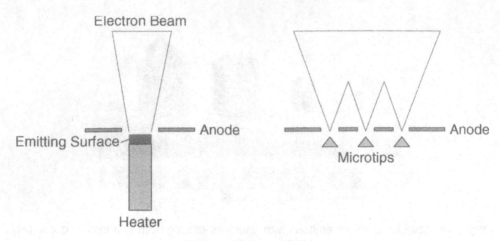

Fig. 5.21. Low power neutralizers

contamination concern. Specifically alkali metals such as Cesium and Rubidium react violently with water and oxygen. Special sealing of the propellant prior to launch and proper handling during operation are very important.

The specific power of field emission thrusters is in the range of 60–75 W/mN. Therefore, especially at micro-Newton thrust levels, the neutralizer must operate at very low power levels (order of Watts) to keep the power consumption at an acceptable level. This limits possible solutions to basically three different solutions: thermionic, field emission array (FEA) cathodes, and carbon nanotube electron sources (see Fig. 5.21).

For reasons of simplicity, both Cs-FEEP and In-FEEP thrusters are currently equipped with thermionic neutralizers based on low-work function impregnated cathodes (e.g. Barium oxide impregnated cathode). The cathode is heated, and an acceleration potential is applied to extract the electrons and inject them into the ion beam. Total power consumption is about 0.4 W/mA, suited for a 100 μN thruster.

As an alternative, field emission electron sources do not require any heating power and are therefore very attractive neutralizer candidates. They consist of either microtips or carbon nanotubes and a micromachined acceleration electrode. If a suffiently high potential is applied between both electrodes, electrons are released and an electron beam is generated. As an example, Fig. 5.22 shows such a FEA cathode close-up built by SRI. However, FEA cathodes have not been lifetime-tested for typical LEO orbits, but are already mature enough for GEO and interplanetary orbit missions. Carbon nanotubes seem to be more robust, but are less efficient than FEA cathodes and have not been tested long enough. After more development, both cathodes will be strong candidates to replace thermionic cathodes in the near future.

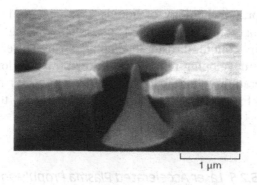

1 µm

Fig. 5.22. FEA cathode close-up (Courtesy of SRI)

5.2.4 Colloid Thruster

A colloid thruster electrostatically accelerates very fine droplets of an electrically-charged, conducting fluid. Droplets are formed by having the liquid flow through a needle with an inner diameter of the order of hundreds of microns. The needle is on positive potential with respect to an accelerator electrode (in some designs on negative potential) similar to field emission thrusters. At the needle exit, a liquid cone developes due to the equilibrium of electrostatic forces and surface tension (Taylor cone). The same electrostatic forces break off droplets from the tip of the cone with a net charge and accelerate it. Several needles are typically clustered together in an array to achieve higher thrust levels (see Fig. 5.23). Such type of thrusters were extensively studied in the 1960s through the mid 1970s in the US (NASA, TRW, US Air Force), Europe (ESA) and Russia (MAI). They were originally oriented as an alternative to ion engines at mN thrust levels operating at a lower specific impulse. The high mass-to-charge ratio of a charged droplet with respect to single ions was known to increase the thrust density and efficiency. On the other hand, very high acceleration voltages were required in the order of 12 up to 100 kV, a requirement very difficult to handle on a spacecraft. Moreover degradation problems due to radiation with the organic propellants used stopped the development.

Fig. 5.23. Colloid thruster (Courtesy of Busek)

However, electrospray technology of liquids have advanced considerably since then, and microspacecraft as well as precision-pointing propulsion needs renewed interest in colloid thruster technology. New research programs are presently aiming at developing miniaturized, low-power colloid thrusters with a much lower acceleration voltage below 10 kV. A low-power neutralizer, similar to those used for Field Emission Thrusters, is necessary to maintain charge neutrality.

5.2.5 Laser Accelerated Plasma Propulsion

A very advanced concept is presently under investigation at the University of Michigan in the United States using lasers to produce very high electrostatic fields. If an ultrashort laser (picoseconds and below) with very high intensities (hundreds of Terrawatts) is focused on a sufficiently small spot, a high temperature plasma is created at the surface of the target, producing electrons at relativistic speeds close to the speed of light. These electrons penetrate the target generating very high electric fields (GV/m), which in turn can accelerate ions out of the target material (see Fig. 5.24).

Using state-of-the-art lasers (500 J with a 500 fs pulse length), the performance of a thruster based on laser acceleration would be a thrust of 100 mN at a specific impulse of 10^7 s! This is a huge velocity at thrusts comparable to standard Ion and Hall thrusters. Of course, the major obstacle is to put such a laser system into orbit and to operate it, requiring a nuclear power plant which delivers 1 MW in this example. However, projections of laser performance in the future suggest that even shorter laser impulses are possible at higher repetition rates leading to thrusts of several tens of Newtons at relativistic specific impulses which is definitely worthwhile investigating and would be a true enabling technology for interstellar missions.

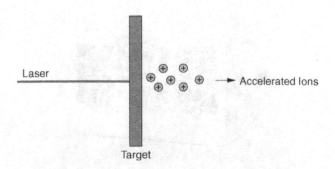

Fig. 5.24. Laser accelerated plasma propulsion concept

5.3 Electromagnetic

Electromagnetic thrusters accelerate ionized propellant through either self-induced and/or applied magnetic fields. The force is expressed by the Lorentz law (see Eq. (3.6)). The high electric currents needed to generate thrusts require high power levels (tens of kW up to MW) or pulsed-mode operation with heavy capacitors.

5.3.1 Magnetoplasmadynamic (MPD) Thruster

In a magnetoplasmadynamic (MPD) thruster or Lorentz force accelerator (LFA), a high current discharge (kilo Ampére at <100 Volts) is applied between two coaxial electrodes (see Fig. 5.25). This discharge ionizes the propellant flowing through the electrode configuration. A variety of non-oxidizing propellants (oxides lead to cathode degradation) can be used such as hydrogen, hydrazine, Argon and even alkali metals like Lithium. The current generates a self-induced radial magnetic field that interacts with the currents, resulting in a Lorentz force acceleration (thrust scales with I^2). The axial component generates thrust and the radial component increases the pressure towards the center line. Propellant flow and thermal gas expansion (temperatures higher than 2,500° C) also contribute to the thrust but are usually at least one order of magnitude below magnetic acceleration. This principle is called self-field MPD thruster. An additional external magnetic field can be added (permanent magnets or coils) around the anode to increase performance, which is known as applied-field MPD thruster.

MPD thrusters can generate very high thrusts up to 200 N, and can transmit very high power levels (up to MWs) at specific impulses of up to 8,000 seconds. The higher the power, the better the efficiency, which is typically around 35% and can increase up to 75%. These performance characteristics make MPD thrusters very attractive for manned missions. Obviously, these high power requirements can only be met in combination with nuclear power plants. Also ground testing is very difficult. One of the few test centers with vacuum facilities as well as available

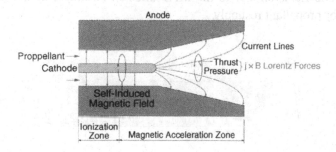

Fig. 5.25. Magnetoplasmadynamic (MPD) thruster

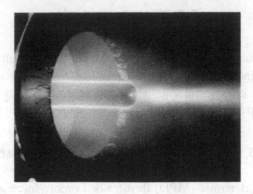

Fig. 5.26. Pulsed MPD thruster (Courtesy of EPPDyL Princeton University)

power for continuous operation of MPD thrusters is the Institute for Space Systems at the University of Stuttgart in Germany. Many other research institutions use an MPD thruster in pulsed operation with large capacitor banks.

Cathode erosion is one of the major lifetime-limitation factors. However, being developed since the 1960's, MPD thrusters have already demonstrated lifetimes in excess of 1,000 hours. A pulsed MPD thruster was successfully flown by the Japanese Institute of Space and Astronautical Science (ISAS) in 1997. Research on this thruster is currently being carried out by NASA JPL, Princeton University, several Russian institutes such as the Moscow Aviation Institute (MAI), European and Japanese laboratories.

5.3.2 Pulsed Plasma Thruster (PPT)

The pulsed plasma thruster is a very simple electric propulsion thruster. A solid propellant bar, usually Teflon®, fills the gap between two electrodes connected to a capacitor bank (see Fig. 5.27). A spark plug triggers an electrical discharge across the surface of the propellant. Heat transfer from the discharge ablates part of the propellant and ionizes it. The ionized propellant is then accelerated by electromagnetic fields similar to the MPD thruster. A simple spring advances the bar axially for propellant resupply.

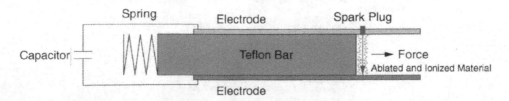

Fig. 5.27. Pulsed plasma thruster (PPT)

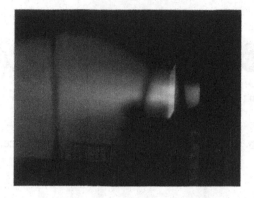

Fig. 5.28. EO-1 flight PPT firing in ground test (Courtesy of General Dynamics-OTS, Space Propulsion Systems)

One shot produces a thrust of tens to hundreds of μN, the thruster usually operates at a frequency of 1–3 Hz. A PPT thruster has several advantages like zero warm-up time and zero standby power, no propellant tanks and feedlines, and most notably is very cheap due to the simple design. The main disadvantage is the very small efficiency between 5–15%.

A PPT thruster was successfully flown on the Soviet Zond 2 spacecraft towards Mars as early as 1964, several applications followed in the United States (see Fig. 5.27). Presently PPT thrusters are under evaluation for fine attitude and position control of formation-flying satellites on account of their small impulse bit capability.

5.3.3 Variable I_{sp} Plasma Rocket (VASIMR)

A plasma rocket uses magnetic fields to confine a plasma and electromagnetic energy to heat it. In the VASIMR concept, hydrogen is ionized and injected into the main chamber by a magnetoplasmadynamic (MPD) device (see Fig. 5.29). The magnetic fields isolate the plasma from the chamber walls so that the temperature can exceed material melting limits and therefore reach very high specific impulses. In the main chamber, electron and ion cyclotron heating is used by applying resonant radiofrequency (RF) fields to heat the plasma, which is then expelled through a magnetic nozzle. Hydrogen propellant can be injected into the plasma exhaust for cooling purposes. The magnetic nozzle can control the thrust by varying the nozzle entrance section. On the other hand, the specific impulse can be varied by changing RF heating power (3,000–30,000 s). This enables to change thrust and specific impulse independently of each other and enables performance to be tailored to a specific mission.

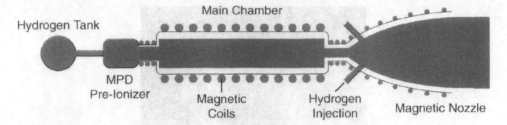

Fig. 5.29. Variable I_{sp} plasma rocket (VASIMR) concept

Work on the VASIMR concept was first conducted at the MIT Plasma Fusion Center in the early 1980's and is presently going on at the Advanced Space Propulsion Laboratory at NASA's Johnson Space Center. A small-scale VASIMR prototype is scheduled to be tested on board the International Space Station to verify performance parameters.

5.4 Induced Spacecraft Interactions

Although electric propulsion systems have been under active development since the 1960's, their application in the field of telecommunication satellites and interplanetary spacecraft, with exception of Russian satellites and technology demonstrations, stagnated until a few years ago. First of all, this was due to the power demands, which were not always available. But secondly, electric propulsion systems (except Resistojets) emit a plasma which can interact with the ambient plasma in space and even more important with the spacecraft itself.

The most important interaction is caused by so-called charge-exchange (CEX) ions. Electric thrusters are not 100% efficient, and therefore emit neutral propellant in addition to ions and electrons. This neutral propellant has very low thermal velocities (hundreds up to thousands of m/s) in comparison with ionized propellant (ten thousands of m/s). If the neutral and ionized propellants collide, charge-exchange collisions can occur and the impulse between both collision partners is exchanged. This leads to fast neutrals and slow ions:

$$\text{Ion}_{Fast} + \text{Neutral}_{Slow} \rightarrow \text{Ion}_{Slow} + \text{Neutral}_{Fast}. \qquad (5.8)$$

These slow charge-exchange ions are strongly affected by the potential distribution around the thruster. Usually, the primary ion beam creates a positive potential in front of the thruster, which repels the charge-exchange ions. Due to beam divergence, the charge-exchange ions can also be pushed back to the spacecraft surface. This is called backflow (see Fig. 5.30).

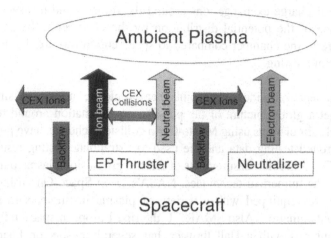

Fig. 5.30. Induced spacecraft interactions

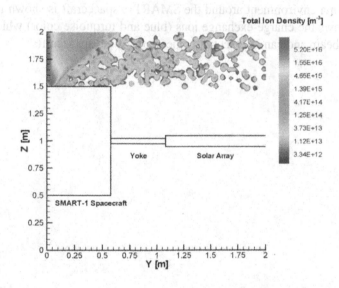

Fig. 5.31. Simulation of SMART-1 hall thruster plasma and charge-exchange ions

Depending on the energy of the charge-exchange ions, this impact can cause sputtering, which is very dangerous, e.g. for the solar arrays, and can degrade the spacecraft's lifetime. They can also interfere with plasma instruments, contaminate optical instruments, cause charging of the spacecraft and frequency shifts of communication waves. These are all very important issues related to safe operation of a spacecraft equipped with electric propulsion, and require careful analysis.

Ground-testing of electric propulsion thrusters in vacuum chambers are one part of this evaluation. Several plasma instruments like Langmuir probes investigate the distribution of charge-exchange ions around the thruster and measure their energy. The influence of the potential distribution by the chamber walls and the relative high pressure in the chamber compared to space conditions are the most important limits to ground testing.

Since instruments provide only point measurements, numerical simulations are needed to get a global picture of the possible contamination around the spacecraft. Many particle simulations using Monte-Carlo collision schemes have proved to match well with ground testing data and are used to extrapolate testing results into space conditions. Few in-orbit measurements of electric propulsion plasma plumes exist that can validate the numerical techniques. NASA's Deep Space One using the NSTAR ion thruster was equipped with a variety of plasma instruments to study charge-exchange contamination. Also SMART-1, the first European spacecraft to the moon, which is equipped with a Hall thruster, has several sensors on board to validate numerical simulations that can be used by satellite manufacturers to avoid contamination and to put thrusters and instruments in safe locations. The Hall thruster induced plasma environment around the SMART-1 spacecraft is shown in Fig. 5.31. It clearly shows the charge-exchange ions (blue and turquoise color) which leave the primary ion beam and can interact with other parts of the spacecraft.

Chapter 6

Micropropulsion

Micropropulsion is a relatively new area involving a drastic reduction of thruster size and mass. This is mainly due to two reasons:

(i) Computer and microstructure manufacturing is making dramatic advances, many spacecraft components such as science payloads (camera, sensor, etc.) or on-board computers can be therefore significantly reduced in size. This has enabled a new generation of microsatellites, which can be launched at much lower costs. Alternatively, many microsatellites can be launched at one time and the risk of failure can be distributed between them. Also new microsatellite formation scenarios e.g. for future Mars missions, are presently under evaluation. All these new spacecraft need propulsion systems that are also reduced in size and mass.

(ii) Micropropulsion technology can reduce the thrust to mass ratio, and in some cases the power to thrust ratio. Therefore many of them in parallel can achieve the same thrust as "normal" propulsion devices, but at a reduced mass and maybe power requirement. This improves the mass ratio in Tsiolkovski's equation, and increases the Δv capability of the spacecraft.

Typical microspacecraft have weights of below 100 kg. New designs require weights of below 10 kg (nanosat) and even below 1 kg (picosat). Such satellites require highly integrated propulsion technologies, which are manufactured using MicroElectroMechanical Systems (MEMS). This chapter can only give a very short overview due to the many diverse developments at present taking place.

6.1 Chemical Propulsion

6.1.1 Solid Microthrusters

Honeywell is working on a digital microthruster, consisting of a multitude of microfabricated, single-shot thrusters placed onto a silicon wafer (see Fig. 6.1). A MEMS thruster array contains a quarter of a million separate thrusters on 3.3×3.3 cm silicon square. Each one has its own heater element individually addressable and ignitable. The MEMS mega-pixel microthruster is following a

Fig. 6.1. Honeywell/Princeton mega-pixel digital microthruster array (Courtesy of Honeywell, Princeton University)

two-stage approach: In the first stage, a small (approx. 1 nanogramm) thermally-detonatable piece of lead styphnate is heated to its auto-ignition temperature (270°C). This styphnate explodes, releasing a lot of energy but non-usable momentum. In the second stage, the exploding styphnate ignites a special nitrocellulose mixture (0.5–8 µg), which then creates thrust. The components are chosen in order to instantaneously convert all propellant to gas without any burn-products. Of course, one of the main challenges is that one explosion shall not damage or trigger any other pixel thruster. Also the propellant filling process is quite complicated. The total mass of the pixel array including fuel is only 2.4 gram, I_{sp} is estimated between 100–300 s and the impulse per pixel can be adjusted from 0.5–20 µNs. Only 10 mW of power is required to ignite a pixel.

There are a number of similar developments going on, most notably a common project between TRW, Aerospace Cooperation and CALTECH sponsored by DARPA (US Defense Advanced Research Projects Agency), and a micro-pyrotechnics project at CNRS in France.

6.1.2 Micro Mono-propellant Thruster

The NASA Goddard Spaceflight Center is currently developing a mono-propellant thruster to be used on nanosatellites. Using MEMS technology, the performance goals are a thrust of 10–500 µN at an I_{sp} of 240 s and an impulse bit of 1–1000 µNs. One major investigation is the fuel choice, which will have a freezing point so low that no heaters are required, thus simplifying the design. Initial work together with

PRIMEX suggests a mixture of Hydrazine (N_2H_4) with HN and H_2O with a freezing point of $-2°C$.

6.1.3 Micro Bi-propellant Thruster

Microfabricated bi-propellant thrusters are currently under study at MIT in the United States and at Mechatronic in Austria. The goal is to integrate all thruster components on a single chip. The MIT design aims at thrust values up to 15 N with mass flow rates of 5 g/s.

6.1.4 Cold Gas Thruster

Cold gas thrusters are a very simple form of rocket technology using a pressurized gas (usually Nitrogen) fed through a valve and a nozzle. Valve leakage is one of the most critical issues, loss rates of 10% of the propellant throughout a mission are not unusual. The smallest thruster available is built by Moog (see Fig. 6.2), with a maximum thrust of 4.5 mN and a tank pressure of 34.5 kPa. The pressure directly corresponds to a specific impulse of 65 s. In lower thrust ranges, the specific impulse will drop due to the pressure drop from the valve to the nozzle. The total weight of this thruster is 7.34 g.

MEMS-based cold gas thrusters are currently under development at the Angstrom Space Technology Centre at Uppsala University in Sweden. A module of four cold gas thrusters with integrated piezoelectric valve assemblies is machined into a silicon wafer. Thrust values of 100 μN are targeted. The entire thruster assembly including electronics and housing is estimated to weight about 70 g. A similar system is also being investigated at Polyflex in the UK and at Mechatronic in Austria (see Fig. 6.3).

Fig. 6.2. 4.5 mN cold gas thruster (Courtesy of MOOG)

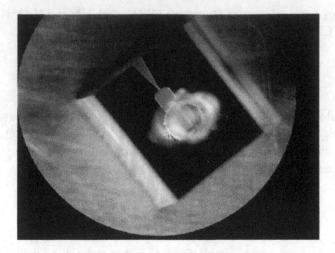

Fig. 6.3. MEMS micronozzle (Courtesy of Mechatronic)

6.2 Electric Propulsion

6.2.1 Micro Ion Thruster

Miniaturized Kaufman ion thrusters with diameters in the 1–3 cm range are currently under development at the University of Southern California together with NASA JPL (see Fig. 6.4). A thrust range of a few mN is targeted. Since a hollow cathode requires a relatively high heating power for proper operation, field emission array electron sources are under investigation for both the electron source inside the discharge chamber and as a neutralizer. Another scaling problem is the magnetic confinement: the smaller the discharge chamber diameter, the higher the magnetic field strengths required for proper confinement.

6.2.2 Low Power Hall Thruster

A 4 mm diameter Hall thruster is under development at MIT in the United States (see Fig. 6.5). The maximum thrust is 1.8 mN at a specific impulse of 865 s, with a power consumption of 126 W and a very poor efficiency of only 6%. Also here, the most difficult scaling problem is the magnetic field strength required to confine the electrons. State-of-the-art Cobalt-Samarium permanent magnets deliver about 1 Tesla at their surface, which limits the size of the diameter for further miniaturization. The magnetic confinement problems are mostly responsible for the poor ionization efficiency.

Fig. 6.4. Micro ion-engine (Courtesy of JPL)

Fig. 6.5. Micro hall thruster (Courtesy of MIT)

Several 50 Watt Hall thrusters are currently being investigated at Busek in the United States, and at the Keldysh Research Center in Russia aiming at higher efficiencies. For example, Busek demonstrated a 100 Watt Hall thruster (although not as miniaturized as MIT's design) with a thrust of 4 mN and an efficiency of 20%.

6.2.3 Micro PPT Thruster

A μ-PPT thruster is currently under development at the Air Force Research Laboratories (AFRL) in the United States. In this design, the Teflon® propellant is put into a coaxial electrode configuration comparable in size to a standard TV cable. The complete thruster including the electronics weighs 0.5 kg; thrust levels are between 2 and 30 μN with a power consumption of 1 to 20 W. During long term testing, the thruster already demonstrated over 500,000 firings.

6.2.4 MEMS FEEP/Colloid Thruster

A Colloid thruster uses a capillary tube for propellant feeding, FEEP thrusters (despite the slit emitter design) use either capillaries or needles. MEMS-based field emission array (FEA) cathode electron sources use highly miniaturized needles and extractor electrode geometries very similar to those used by needle type FEEP thrusters. Even miniaturized volcano (similar to capillary tubes, see Fig. 6.6) structures could be produced out of silicon and integrated with FEA extractor electrodes. Such emitter structures are currently under investigation for use in FEEP and Colloid thrusters. The great advantage is that, because the extractor electrode is only a few μm apart from the emitter instead of half a millimeter, much lower voltages are needed to reach an electric field sufficient to extract ions or colloids. Together with an additional acceleration electrode, this enables the power-to-thrust ratio to be scaled, specific impulse and thrust to be custom-tailored to mission requirements. Major difficulties are presently material compatibility of propellant with the silicon structure and possible overflows of the propellant out of the

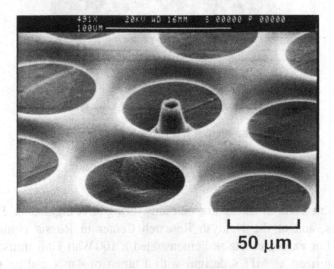

Fig. 6.6. Micro capillary structure (Courtesy of SRI International)

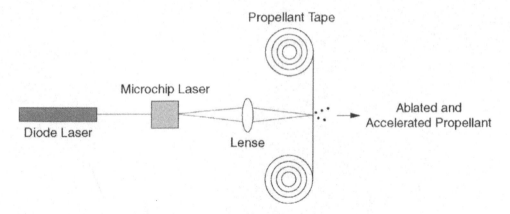

Fig. 6.7. Microchip laser thruster concept

capillary. Several MEMS Colloid thrusters are presently under development in the United States.

6.2.5 Microchip Laser Thruster

A very interesting concept based on microchip laser propulsion is currently being experimentally explored at MIT in the United States. A standard 1 W diode laser is used to pump a microchip laser invested by MIT which transforms the continuous laser light into high intensity pulsed laser light. Using an optical arrangement of lenses, the pulsed laser light is then focused on a propellant tape (e.g., 1s pulse on Teflon® coated with Aluminum), which is heated and emits ablated propellant material at high thermal velocities (see Fig. 6.7). The complete thruster is simple, weighs only 400 g including the PCU, and is able to provide thrust ranges from 0.3 nN up to 3 μN at a power consumption of 6.5 W. The specific impulse is in the range of 1,000 s. The big advantage of the pulsed laser light is the very high intensity (in preliminary experiments intensities of 20 GW/cm² were achieved) and the low temperature conductivity losses on the propellant tape. Work is in progress to optimize this thruster for higher efficiencies (presently around 1%) and higher thrust levels.

Chapter 7

Propellantless Propulsion

Some advanced propulsion systems do not require any propellant at all (or at least very little in the case of Mini-Magnetosphere propulsion), which makes them very attractive for several applications such as satellite de-orbiting or even interstellar missions. The energy required can be either supplied internally (solar arrays, on-board nuclear power generation) or externally (laser, microwave, sun, magnetic fields). All concepts presented are still in the demonstration/feasibility concept phase but soon will be used quite frequently such as tethers.

7.1 Tethers

Tethers are long cables connected to a spacecraft, which can provide orbit raising/de-orbiting maneuvers as well as propulsion or power generation through the Earth's (or other planet's) magnetic field. We distinguish in the following between momentum exchange tethers and electrodynamic tethers that interact with such magnetic fields.

7.1.1 Momentum Exchange Tether

If a long tether is spinning around an orbit, it can pick up payload at a low orbit and spin it up to a higher orbit. This is called a "bolo" (see Fig. 7.1). An extreme version is the rotavator where the cable is extended to reach the planet's surface. Atmospheric drag will obviously slow down and heat up such a tether significantly so that no material presently exists for any Earth applications. However, such momentum exchange tethers could be constructed for the Moon or Mars, which have a much thinner atmosphere and therefore much less drag.

7.1.2 Electrodynamic Tether

The most commonly known concept is the electrodynamic tether. Here, a long cable consisting of a conductive core and an insulator coating interacts with a planet's magnetic field (e.g., from the Earth) to generate power or provide propulsion. One end of the tether needs to be mounted on the spacecraft structure. Depending on the operation mode, an electron emitter (plasma contactors, field emission array

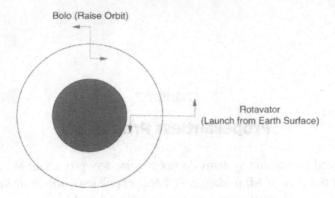

Fig. 7.1. Momentum exchange tether

cathodes, etc. similar to electric propulsion neutralizers described in chap. 5) needs to be mounted at the other end of the tether or on the spacecraft in order to emit a current into the ambient space plasma interacting with the planet's magnetic field if the spacecraft is moving perpendicular to it. The maximum force can be expressed by

$$F = \frac{B^2 l^2 v}{R},$$ (7.1)

where B is the magnetic field (for the Earth in LEO $B = 20\,\mu T$), l the length of the tether, R its resistance and v the spacecraft's velocity. For a length of 5 km, the resistance of an aluminium tether is about 185 Ω. For a typical LEO velocity of 6,800 m/s, the tether will therefore produce a maximum force of 0.36 N. Operating in reversed mode, the tether will generate a power of

$$P = F \cdot v = \frac{(B \cdot l \cdot v)^2}{R}.$$ (7.2)

In our example, this would yield 2.4 kW. Of course in this case, a drag force would apply to the tether and slow down the spacecraft. This can be used for de-orbiting a satellite. This power would need to be applied to the tether to produce the 0.36 N of thrust. This is not small and could, on a larger scale, be used, e.g. to continuously reboost the international space station.

Several flight demonstration tests of a tether system demonstrated the principal capabilities of an electrodynamic tether. Obviously the most difficult part is the deployment of a several kilometer-long tether cable. The longest tether ever successfully deployed had a length of 20 km. High voltage drops along the cable and arcing are another problem. However, due to the principal simplicity of the concept and the advantage of no propellant being required promise a bright future for tether propulsion.

7.2 Propellantless Electric/Nuclear Propulsion

Electric and nuclear propulsion mostly use Xenon or Hydrogen gas stored in big tanks on board the spacecraft. However, the propellant gas does not necessarily need to be on board the spacecraft. One can also envision collecting the neutral gas atmosphere close to a planet (like in LEO orbit or in a low Marsian orbit) and then utilizing it as a propellant. In the case of LEO orbit, oxygen is predominantly available. Recent studies estimate a realistic power to thrust ratio of 80 W/mN, which is up to 25% higher than the one for Field Emission Thrusters.

Another very advanced concept is called the interstellar ramjet, collecting the interstellar hydrogen gas which penetrates the galaxies, using it as a fuel for a fusion reactor and creating thrust by expelling the heated hydrogen gas through a nozzle. It is estimated that the collecting area needs to be about 10,000 km^2 for an acceleration of $10\,\mathrm{ms}^{-2}$ (equal to the Earth's standard gravitational acceleration) throughout the interstellar medium. If it can be realized, it would be a very attractive solution since it is independent of an energy source AND propellant.

7.3 Photon Rocket

A very simple concept is to directly convert electric energy into kinetic energy via the use of a laser. Photons are then used as a propellant producing thrust. The energy stored in a laser is proportional to the frequency of the emitted light f and Planck's constant h:

$$W = h \cdot f. \tag{7.3}$$

Since a photon's velocity is the speed of light c, thrust and specific impulse for a photon rocket can be expressed by

$$F = \frac{P}{c} = \frac{h \cdot f \cdot R}{c}, $$

$$I_{\mathrm{sp}} = \frac{c}{g}, \tag{7.4}$$

where P is the power and R the repetition rate. Using a 1 MW laser, we will get a thrust of 3.3 mN and a specific impulse of 3×10^7 s. Of course, the concept of a specific impulse does not make sense any more since we do not "consume" propellant out of a tank reservoir, and Tsiolkovski's equation no longer applies. This is also a propulsion concept which does not have limits on the achievable maximum Δv as long as there is electric energy available. The power-to-thrust ratio is enormously high, nevertheless, it is a true propellantless propulsion system which

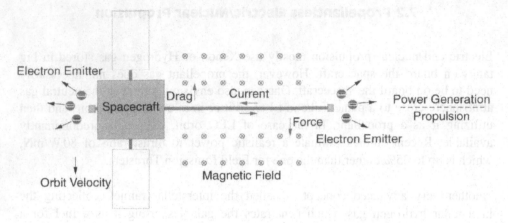

Fig. 7.2. Electrodynamic tether

works independently of any orbit. If new breakthroughs in on-board energy production are made (e.g., fusion power plants), maybe such a propulsion system can enable interstellar propulsion capabilities.

7.4 Beamed Energy Earth-to-orbit Propulsion

Lasers can be used to heat air and create thrust. Figure 7.3 shows a small demonstration vehicle ("Laser Lightcraft"), which focuses laser light on ambient air and uses a plug nozzle (see chap. 2.3.2) to create thrust. The power necessary

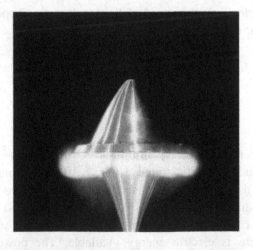

Fig. 7.3. Laser lightcraft (Courtesy of NASA)

to launch a vehicle with beamed energy can be estimated by at least a MW per kilogram. Therefore, very large power-level lasers are required (or alternatively only very small launch masses), which limits the application of this propulsion concept. The US Air Force is presently experimentally investigating this concept using pulsed MW lasers and has already achieved a height of 100 m for a kg class vehicle. Also the German Aerospace Center DLR is investigating such a concept.

7.5 Solar Sails

A very popular propellantless propulsion concept is the solar sail. Here, the pressure from the solar photons are collected using a very large sail producing thrust. The solar pressure at the Earth's orbit is around 9 N/km^2 decreasing with the $1/r^2$ from the sun. Therefore, especially for interplanetary missions, very large sail structures are required for useful thrusts. In addition to a main sail, small steering sails (typically 5% in size of the main sail) are attached at the ends of a square solar sail for attitude control purposes (see Fig. 7.4). Obviously, the main critical points are the deployment mechanism and the mass of the sail structure. Sail materials presently under study have a thickness of only 1 μm which is the limit for the material to be able to survive the deployment and launch period. This then gives a thrust-to-weight ratio of about 10^{-5} N/kg. Space-manufactured solar sails may not require deployment mechanisms and do

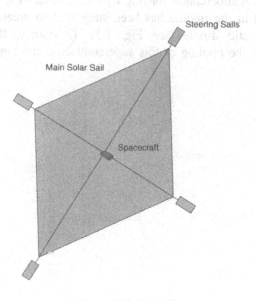

Fig. 7.4. Solar sail

not have to withstand launch vibrations, which might improve this ratio with even thinner materials.

Solar sails can increase and decrease orbital speed and therefore can be used to go either close to the sun or to the outer solar system. Also maneuverability is good, comparable to a windsurfer who can maneuver around the sea as well with only one wind direction. The basic concept can be improved by separating the function of collecting photons and reflecting the photons to create thrust. We would then have one sail always facing the sun so that the collecting area is always maximized. This sail is slightly curved so that the reflected photons are concentrated and therefore can be collected by a smaller second sail, which reflects them in the desired direction.

7.6 Magnetic Sails

The solar wind travels at speeds ranging from 300–800 km/s, which is very high compared to the fastest man-made spacecraft Voyager, with a present speed of about 17 km/s. Since the solar wind consists of charged particles, a magnetic dipole can deflect the solar wind and therefore create thrust. The main drawback is that it has a very low density, which requires either very high field strengths or very large magnetic fields to create useful thrust for standard-size spacecraft.

Presently, there are two different concepts:

(i) Magnetic sail: A superconductor ring with a diameter of several kilometers and the spacecraft in the middle has been suggested to create a large high field strength magnetic dipole (see Fig. 7.5). Obviously, the major technical difficulties are the cooling of this superconductor, the large structure and the weight.

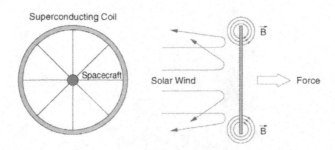

Fig. 7.5. Magnetic sail

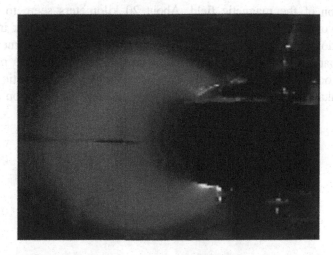

Fig. 7.6. M2P2 prototype (Courtesy of Robert Winglee, University of Washington)

(ii) Mini-magnetosphere (M2P2): A very elegant way was recently proposed whereby a large-scale magnetic field would be produced by a magnetic dipole on board the spacecraft and a plasma (e.g., Argon etc.) produced by a Helicon device (radiofrequency plasma generator using a coil, see Fig. 7.6) would be injected. Since the plasma increases the magnetic permeability, the magnetic field expands with the plasma to a few kilometers in diameter (see Fig. 7.7). Presently, modeling and experimental efforts are underway to estimate the

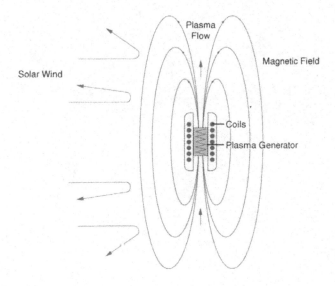

Fig. 7.7. Mini-magnetosphere propulsion

expansion of the magnetic field. About 20 kilometers seem to be possible, which would be large enough to produce significant thrusts for interplanetary exploration missions. Since some gas is used to produce the magnetic field, it is not primarily used as a propellant – and thus M2P2 is listed under propellantless propulsion. In principle, the plasma is contained in the magnetic bubble, and only leakage needs to be replaced. So "propellant" consumption will be very small.

Chapter 8

Breakthrough Propulsion

8.1 Current Fundamental Limitations in Propulsion

Although this book discusses many advanced concepts that enhance present propulsion systems, in nearly every aspect we are already at (or know) the limit. For example, the LO_2/LH_2 chemical rocket engine from the space shuttle orbiter has a combustion efficiency of more than 99%. That does not leave lots of room for improvement. Electric propulsion systems combined with nuclear power generators or pure nuclear rockets like those developed in the NERVA program will enable us to explore our solar system with people instead of robots within reasonable amounts of time. This represents already a big step forward from where we are now (we were going to the moon in a 3 day trip). However, this is all VERY far away from a real revolution in space travel, namely exploring OTHER solar systems within a crew's lifetime!

How far are we from doing this? Well, the closest star to our own sun is *Proxima Centauri,* part of a three-star system called Alpha Centauri, which is about 4.3 lightyears (the distance light travels in one year $\cong$ 10 million million kilometers) away. According to Table 1.1 in chap. 1, the Δv required for such a distance to be achieved in a reasonable amount of time (10 years) is about 100,000 km/s. Today's robotic planetary exploration missions require about 10 km/s. Therefore, we are roughly about 4 orders of magnitude away from achieving manned interstellar spaceflight (we don't know if Alpha Centaury has any solar system at all; maybe we will have to go on an even longer trip to find one). Even with high performance nuclear rockets, we are still more than 2 orders of magnitude away from this goal.

What are the fundamental limits that we face in our technology? In addition to Newton's mechanics, two theories in physics describe these limits best:

(i) *Thermodynamics* (chemical rockets, nuclear thermal, etc.): Energy can only be transformed; it cannot be created out of nowhere (no Perpetuum Mobile).

(ii) *Relativity Theory:* The fastest possible speed in vacuum is the speed of light (fundamental speed limit); mass is a function of spacetime curvature (we need something like a black hole to modify space, time and mass).

That does not sound very promising if we are looking for a propulsion system that is very fast, compact in size and weight, and of course does not require huge amounts of energy! However, new theories (or old ones which were not pursued in the past) are emerging questioning all these limitations. In fact, there is a good chance based on historic facts that there will be a breakthrough: Around the beginning of the 20th century, scientists "proved" that it is impossible to fly with a machine that is heavier than air; in the 1930s, scientists "proved" that it is impossible to fly to the moon (the energy required would be enormous), and last but not least, scientists "prove" today that it is impossible to go to another star. So in principle, all you have to do is to wait long enough until the technology is ready. Or you can actively try to stimulate such a breakthrough.

In 1996, NASA established the Breakthrough Propulsion Physics Program to seek for breakthroughs in propulsion systems that require no propellant, need as little power as possible and allow the fastest possible speeds. This program already sponsored research which is currently underway evaluating claims in the literature, new coupling phenomena between gravitation and electromagnetism as well as predictions based on Quantum theory. In addition, other government agencies like the US Department of Energy or the European Space Agency have started to evaluate possible benefits from such kinds of exploratory research. This chapter will describe some of the concepts that are seeking for breakthroughs and how they could change the way we at present access space.

8.2 Quantum Physics, Relativity Theory, Electromagnetism and Space Propulsion

Three theories have played a key role in 20th century physics: Quantum theory and all its consequences for nuclear physics; relativity theory describing space, time and cosmology, and of course electromagnetism describing electricity, on which basically all technology depends. How can we use these theories, combine them or find new relationships which would benefit space propulsion?

Let us explore two examples:

8.2.1 Quantum Theory and the Casimir Effect

As outlined above, the fastest possible speed is the speed of light in vacuum. What if we could modify the vacuum itself? In fact, Quantum theory even describes how we can do it. One of its most fundamental laws is the Heisenberg uncertainty principle, which states that it is not possible to measure the exact position and the exact impulse of a particle at the same time. Any atom vibrates at a certain frequency f according to its temperature T. But what happens if the temperature is zero Kelvin? In classical mechanics, the atom would then stand still, and its position as well as its impulse ($\equiv$ zero) could then be determined exactly. However, this violates the Heisenberg uncertainty principle, and therefore Quantum theory says that the atom is still moving even at zero Kelvin temperature. The average energy (we don't know the exact value in Quantum theory) is called Zero Point Energy (ZPE) given by

$$\langle E \rangle_{\text{ZPE}} = \frac{hf}{2},$$ (8.1)

where h is Planck's constant. This energy penetrates through the whole universe and is always present independent of temperature. What about the vacuum? The vacuum is not "empty" since it has dielectric properties and can support radiation (like electromagnetic waves). Moreover, the ZPE provides a source of energy for the vacuum. In principle, all frequencies, and correspondingly all wavelengths, are possible for the ZPE.

If the vacuum is now limited, e.g. by two conducting parallel surfaces to shortcut, e.g. the wavelength, the possible wavelengths (and hence frequencies) between them cannot take any value any more but are limited by the distance L of the two surfaces. Hence, the vacuum is less energetic perpendicular to the surfaces than parallel to them. The higher energy vacuum outside the surfaces presses them together with a pressure of

$$\frac{F}{A} = \frac{\pi h c}{480 L^4}.$$ (8.2)

This force was predicted by Henrik Casimir from Phillips Research Laboratories in 1948, and is called the Casimir effect. The force on two parallel conducting plates with an area of $1\,\text{m}^2$ at a distance of $1\,\mu\text{m}$ is therefore $1\,\text{mN}$. Recent accurate force measurements verify Casimir's predictions within 1%. Obviously, if we would engineer the vacuum using micro and nanoscale structures in order to create a force as

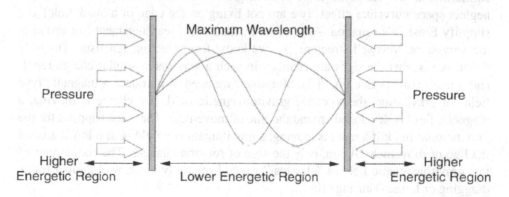

Fig. 8.1. Casimir effect

outlined above, we are far away from Newton's mechanics and the classical reaction principle. Moreover since we modified the vacuum structure, the vacuum speed of light will change as well. As predicted by Scharnhorst in 1990, the difference in the speed of light perpendicular to the surfaces and parallel to the surfaces (where the vacuum is undisturbed) can be expressed as

$$\frac{c_\perp}{c_\parallel} = 1 + \frac{1.59 \times 10^{-56}}{L^4}. \qquad (8.3)$$

Hence, the speed of light will be higher along the lower energetic vacuum region (although only marginal). This ZPE may also provide a source of presently unexploited energy that could power an interstellar spacecraft. Also inertial mass might be modified by vacuum engineering due to the change of the speed of light ($E = mc^2$). Other theoretical concepts try to explain the inertial mass as a Lorentz force interaction of particles moving in a ZPE background field (dating back to ideas of Sacharov in the 1960s, the inventor of the Russian nuclear fusion bomb).

As we can see, Quantum theory can provide some new ways to follow that might lead to a breakthrough in space propulsion. Several of the outlined possibilities are currently being explored in the United States and experiments are under way to measure some of the predictions.

8.2.2 Coupling of Gravitation and Electromagnetism

Contemporary technology is mainly based on electricity. We have learned how to generate electric fields, amplify and modify them to custom needs. A similar technology for gravitation would certainly be very attractive for both space travel and microgravity research on Earth. So what is the difference between electromagnetism and gravitation? When we compare Maxwell's equations for electromagnetism and Einstein's general relativity theory, we hardly see any similarities at all. However, if we look at gravitation at our energy scale, we can neglect space curvature effects (we are not living on the edge of a black hole) and simplify Einstein's equation. After some mathematical rearrangement, we arrive at the very same Maxwell structure that we know from electromagnetism. The only difference is, that the signs are changed in such a way that a similar charge repels and a mass attracts (see Fig. 8.2). Moreover, we need to introduce a magnetic type field for gravitation, the so-called gravitomagnetic field. If a charge is moving, a magnetic field is developed around the line of movement. The same happens for the gravitational field: if a mass is moving, a gravitomagnetic field is developed around the line of movement. Again, only the sign of rotation changes. The counterpart of this gravitomagnetic field in "classical" general relativity is the well-known frame dragging or Lense-Thirring effect.

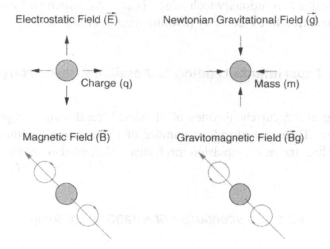

Fig. 8.2. Similarity between electromagnetic and gravity-gravitomagnetic fields

Newtonian gravity assumes instantaneous interaction, which violates energy and momentum conservation. If gravity is assumed to propagate with the speed of light, the extension to Newtonian gravity is just this gravitomagnetic field, which also falls out of simplifying general relativity theory. Now that we know more about the nature of gravitation at our scale and its similarity to electromagnetism, there exists indeed a mechanism to induce and amplify non-Newtonian (artificial) gravitational fields: it is the simple Faraday's law of induction. The very same technique used to generate electric fields (time varying magnetic fields) now applies to gravitation (time varying gravitomagnetic fields) as well. How do we create gravitomagnetic fields? Exactly this is the challenge. Since both electromagnetism and gravitation originate from the same source, the particle, magnetic fields correspond to gravitomagnetic fields ($\vec{B}_g = \kappa \cdot \vec{B}$) and electrostatic to Newtonian-gravitational fields ($\vec{g} = \kappa \cdot \vec{E}$). The coupling constant between both fields is given by

$$\kappa = -7.41 \times 10^{-21} \cdot \frac{m}{q}, \tag{8.4}$$

which is unfortunately very small. So in principle, a normal magnet can be used to induce a non-Newtonian gravitational field. However, with the electron as the source of the magnetic and gravitomagnetic field in this case, the gravitomagnetic field will be 32 orders of magnitude below the magnetic field. Although this is very small, in principle it is possible! This standard coupling between gravitation and electromagnetism is very small, but possibly amplification mechanisms exist that could generate high order of magnitude effects (similar to Ferromagnetism and its amplification to magnetic fields, etc.). If they can be detected, a revolutionary technology is at hand, which will lead to another industrial revolution like electricity revolutionized our past society.

8.3 Experiments Leading to Possible Breakthroughs

After looking at the current theories of physics, breakthroughs in propulsion do seem possible. There are already a number of experiments dealing with gravitational shielding and new propulsion mechanisms. Let us also explore some of the examples:

8.3.1 Superconductor Gravitational Shielding

Probably the best well known experiment was published by Eugine Podkletnov in 1992, concerning a gravitational shielding effect using rotating superconductors. A disk made out of a $YBa_2Cu_3O_{7-x}$ high temperature superconductor material was placed above a toroidal coil magnet. When the temperature was below 70 Kelvin, the material became superconducting and started levitating above the magnet (superconductors act like "magnetic mirrors" in their superconducting phase, this is called the Meissner effect). Then, an AC current was applied to two coils on the side of the superconducting magnet so that it began rotating (see Fig. 8.3). Also the main toroidal coil was connected to an AC current source so that the superconductor started vibrating as a whole. Several samples (metal, plastic, glass, wood, etc.) were placed

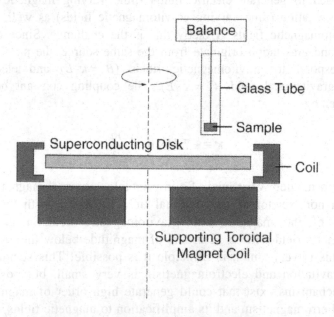

Fig. 8.3. Podkletnov's experimental setup

above the superconducting disk and were found to lose weight with respect to the AC frequency and the temperature. A maximum weight loss of about 2% was recorded. The samples were put inside a glass tube in order to eliminate air flow effects.

A weight loss of this percentage is huge, and obviously far above any effect predicted by general relativity. Moreover, since a superconductor is a macroscopic quantum object, a thorough theoretical explanation requires a quantum theory of gravitation that has not yet been achieved. Podkletnov has, up to 2001, continually published results with modified setups, more precise measurement techniques and improvements of the superconducting material. However, no one else has, up to now, published similar results, although several groups, including NASA, have been trying to replicate them. Unfortunately, none of the groups replicated the whole setup (e.g., static magnetic field and not AC, etc.) completely.

8.3.2 Coupling of Charge, Mass and Acceleration

More than 130 years ago, Michael Faraday carried out experiments based on the idea that electromagnetic and gravitational fields might be inductively coupled. In fact, as we have already seen in this chapter, the fields are indeed coupled by a very small constant depending on the source charge to mass ratio. However, a coupling between charge and mass of the source itself is up to now unknown (but might exist). There are some experiments recorded in the literature, however, that suggest a coupling between charge and mass in combination with rotation (or acceleration, movement in general).

James Woodward, from the University of California at Fullerton, reported experiments on accelerating masses that charged up according to their mass and speed of rotation. His experiments were done both for rotating masses as well as for linear accelerated test bodies. He suggested a broader conservation principle including mass, charge and energy. His results follow:

$$q' \simeq \text{constant} \cdot m \cdot a, \qquad (8.5)$$

where q' is the induced charge, m the test mass and a the acceleration (from rotation or calculated from the impact velocity).

A combination of all these factors to reduce/increase the weight of a body is described in a patent by Yamashita and Toyama (European Patent EP 0 486 243 A2). A cylinder was rotated and charged using a Van der Graff generator. During operation, the weight of the rotating cylinder was monitored on a scale. The setup is shown in Fig. 8.4. If the cylinder was charged positively, a positive change of weight up to 4 grams at top speed was indicated. The same charge negative produced a reduction of weight of about 11 grams (out of 1300 grams total weight). This is a similar relationship between charge, rotation and mass reported by

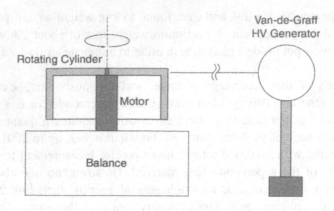

Fig. 8.4. Yamashita et al.'s experimental setup

Woodward. The experimentors note that the weight changed according to the speed of the cylinder, ruling out electrostatic forces, and that it did not depend on the orientation of rotation, ruling out magnetic forces. The reported change of weight (below 1%) is significant and indicates a very high order of magnitude effect.

Experiments need to be conducted to confirm or dismiss these results. The literature is full of similar claims – even if 99% of them are wrong, maybe some of them can lead to a breakthrough in propulsion.

8.4 When Will We Revolutionize Space Travel?

Although rocket propulsion is some 1,000 years old, it was not until 60 years ago that sudden huge advancements lead to the precursor of all modern rocket systems, the V-2. This technology level stayed more or less the same until the end of the 1950's. And then, within about 10 years, the United States built the Saturn-V rocket and sent men to the moon. The space shuttle which was built after the successful moon program in the 1970's is still in service. It is expected that the shuttles will continue flying till 2020. New propulsion concepts such as air breathing rocket propulsion, electric propulsion, etc. are being studied and slowly implemented. Space business is very conservative and prefers to rely on well-tested solutions.

But as this book has shown, there are many revolutionary concepts at hand that will enable manned planetary explorations and even interstellar missions. And there are even more fantastic ideas around that will probably enable us to master gravitation itself and revolutionize space travel. And when will that happen?

As history has taught us many times, it can happen very quickly and within a very short space of time . . .

Further Reading

The following references are recommended for further reading:

General Information on Advanced Space Propulsion Systems

NASA JPL Advanced Propulsion Concepts Notebook Online, *http://sec353.jpl.nasa.gov/apc/*

Encyclopedia Astronautica, *http://www.astronautix.com/*

Propulsion Fundamentals, Chemical Propulsion Systems

Caporicci, M (2000) *Future Launcher Perspectives at the European Space Agency-ESA.* Air & Space Europe, Vol. 2, No. 2

Oberth, H (1923) *Die Rakete zu den Planetenräumen.* München: Oldenbourg

Sutton, GP, Biblarz, O (2001) *Rocket Propulsion Elements.* New York: Wiley

Turner, MJL (2000) *Rocket and Spacecraft Propulsion: Principles, Practise and New Developments.* Springer

Launch Assist Technologies

Toro, PGP, Myrabo, LN, Nagamatsu, HT (1998) *Pressure Investigation of the Hypersonic "Directed-Energy Air Spike" Inlet at Mach Number 10 with Arc Power up to 70 kW.* AIAA Paper 98–0991

Unmeel, BM (2001) *MHD Energy Bypass Scramjet Engine: A Progress Report.* Proc. NASA Advanced Space Propulsion Workshop, Huntsville, Alabama, April 3–5, 2001

Nuclear Space Propulsion Systems

Howe, SD, (1985) *Assessment of the Advantages and Feasibility of a Nuclear Rocket for a Manned Mars Mission.* Proc. Manned Mars Mission Workshop, Marshall Space Flight Center, Huntsville, Alabama, June 10–14

Kammash, T (1985) *Fusion Energy for Space Propulsion.* American Institute for Aeronautics and Astronautics

Lewis, RA, McGuire, T, Mitchell, Smith, G (1997) *Production and Trapping of Antimatter for Space Propulsion Applications.* AIAA Paper 97–2793

Electric Propulsion Systems

Auweter-Kurtz, M (1992) *Lichtbogenantriebe für Weltraumaufgaben*. Stuttgart: Teubner

Brophy, JR, Noca, M (1998) *Electric Propulsion for Solar System Exploration*. Journal of Propulsion and Power 14(5): 700–707

Jahn, RG (1968) *Physics of Electric Propulsion*. New York: McGraw-Hill

Martinez-Sanchez, M, Pollard, JE (1998) *Spacecraft Electric Propulsion – An Overview*. Journal of Propulsion and Power 14(5): 688–699

Stuhlinger, E (1964) *Ion Propulsion for Space Flight*. New York: McGraw-Hill

Micropropulsion

Micci, MM, Ketsdever, AD (2000) *Micropropulsion for Small Spacecraft (Progress in Astronautics and Aeronautics, Vol. 187)*. American Institute for Aeronautics and Astronautics

Propellantless Propulsion

Forward, RL (1990) *Solar Photon Thruster*. Journal of Spacecraft 27(4): 411–416

Forward, RL, Hoyt, RP (1998) *Application of the Terminator Tether Electrodynamic Drag Technology to the Deorbit of Constellation Spacecraft*. AIAA Paper 98–3491

Morgan, JA (1999) Neutrino Propulsion for Interstellar Spacecraft. Journal of the British Interplanetary Society 52: 424–428

Myrabo, LN, Messitt, DG, Mead, FB (1998) *Ground and Flight Tests of a Laser Propelled Vehicle*. AIAA Paper 98–1001

Winglee, RM, Ziemba, T, Slough, J, Euripides, P, Gallagher, D (2001) *Laboratory Testing of the Mini-magnetospheric Plasma Propulsion (M2P2) Prototype*. Proc. Space Technology and Applications International Forum (STAIF)

Breakthrough Propulsion

Forward, RL (1996) *Mass Modification Experiment Definition Study (An Air Force Report)*. Journal of Scientific Exploration 10(3): 325–354

Forward, RL (1984) *Extracting Electrical Energy from the Vacuum by Cohesion of Charged Foliated Conductors*. Physical Review B30(4): 1700–1702

Haisch, B, Rueda, A, Puthoff, HE (1997) *Physics of the Zero-point-field: Implications for Inertia, Gravitation and Mass*. Speculations in Science & Technology 20: 99–114

Podkletnov, E, Nieminen, R (1992) *A Possibility of Gravitational Force Shielding by Bulk YBa2Cu3O7-x Superconductor*. Physica C 203: 441–444

Scharnhorst, K (1998) *The Velocities of Light in Modified QED Vacua*. Annalen der Physik (Leipzig), 8. Ser., 7: 700–709

Tajmar, M, de Matos, CJ (2001) *Induction and Amplifications of Non-Newtonian Gravitational Fields*. AIAA Paper 2001–3911

Woodward, J (2000) *Mass Fluctuations, Stationary Forces, and Propellantless Propulsion*. Proceedings of the Space Technology and Applications International Forum (STAIF-2000), edited by EL-Genk, MS. AIP Conf. Proc. 504, American Institute of Physics, New York

Subject Index

SpringerGeosciences

Yong-Qi Chen,
Yuk-Cheung Lee (eds.)

Geographical Data Acquisition

2001. XIV, 265 pages. 167 figures.
Softcover EUR 62,–
(Recommended retail price)
Net-price subject to local VAT.
ISBN 3-211-83472-9

This book is dedicated to the theory and methodology of geographical data acquisition, providing comprehensive coverage ranging from the definition of geo-referencing systems, transformation between these systems to the acquisition of geographical data using different methods. Emphasis is placed on conceptual aspects, and the book is written in a semi-technical style to enhance its readability. After reading this book, readers should have a rather good understanding of the nature of spatial data, the accuracy of spatial data, and the theory behind various data acquisition methodologies. This volume is a text book for GIS students in disciplines such as geography, environmental science, urban and town planning, natural resource management, computing and geomatics (surveying and mapping). Furthermore it is an essential reading for both GIS scientists and practitioners who need some background information on the technical aspects of geographical data acquisition.

SpringerWienNewYork

A-1201 Wien, Sachsenplatz 4–6, P.O. Box 89, Fax +43.1.330 24 26, e-mail: books@springer.at, Internet: www.springer.at
D-69126 Heidelberg, Haberstraße 7, Fax +49.6221.345 229, e-mail: orders@springer.de
USA, Secaucus, NJ 07096-2485, P.O. Box 2485, Fax +1.201.348-4505, e-mail: orders@springer-ny.com
Eastern Book Service, Japan, Tokyo 113, 3–13, Hongo 3-chome, Bunkyo-ku, Fax +81.3.38 18 08 64, e-mail: orders@svt-ebs.co.jp

SpringerGeosciences

Bernhard Hofmann-Wellenhof,
Herbert Lichtenegger,
James Collins

Global Positioning System

Theory and Practice

Fifth, revised edition.
2001. XXIII, 382 pages. 45 figures.
Softcover EUR 51,–
(Recommended retail price)
Net-price subject to local VAT.
ISBN 3-211-83534-2

This new edition accommodates the most recent advances in GPS technology. Updated or new information has been included although the overall structure essentially conforms to the former editions. The textbook explains in comprehensive manner the concepts of GPS as well as the latest applications in surveying and navigation. Description of project planning, observation, and data processing is provided for novice GPS users. Special emphasis is put on the modernization of GPS covering the new signal structure and improvements in the space and the control segment. Furthermore, the augmentation of GPS by satellite-based and ground-based systems leading to future Global Navigation Satellite Systems (GNSS) is discussed.

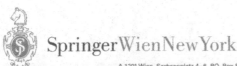
SpringerWienNewYork

A-1201 Wien, Sachsenplatz 4–6, P.O. Box 89, Fax +43.1.330 24 26, e-mail: books@springer.at, Internet: www.springer.at
D-69126 Heidelberg, Haberstraße 7, Fax +49.6221.345-229, e-mail: orders@springer.de
USA, Secaucus, NJ 07096-2485, P.O. Box 2485, Fax +1.201.348-4505, e-mail: orders@springer-ny.com
Eastern Book Service, Japan, Tokyo 113, 3–13, Hongo 3-chome, Bunkyo-ku, Fax +81.3.38 18 08 64, e-mail: orders@svt-ebs.co.jp

SpringerGeosciences

Roman U. Sexl,
Helmuth K. Urbantke

Relativity, Groups, Particles

Special Relativity and Relativistic Symmetry
in Field and Particle Physics

Revised and translated from the German by H. K. Urbantke.
2001. XII, 388 pages. 56 figures and 1 frontispiece.
Softcover EUR 46,80
(Recommended retail price)
Net-price subject to local VAT.
ISBN 3-211-83443-5

This textbook attempts to bridge the gap that exists between the
two levels on which relativistic symmetry is usually presented – the
level of introductory courses on mechanics and electrodynamics
and the level of application in high-energy physics and quantum
field theory: in both cases, too many other topics are more important and hardly leave time for a deepening of the idea of relativistic
symmetry. So after explaining the postulates that lead to the Lorentz
transformation and after going through the main points special relativity has to make in classical mechanics and electrodynamics, the
authors gradually lead the reader up to a more abstract point of
view on relativistic symmetry – always illustrating it by physical
examples – until finally motivating and developing Wigner's classification of the unitary irreducible representations of the inhomoge-
neous Lorentz group. Numerous historical and mathematical asides
contribute to conceptual clarification.

SpringerWienNewYork

A-1201 Wien, Sachsenplatz 4–6, P.O. Box 89, Fax +43.1.330 24 26, e-mail: books@springer.at, Internet: www.springer.at
D-69126 Heidelberg, Haberstraße 7, Fax +49.6221.345-229, e-mail: orders@springer.de
USA, Secaucus, NJ 07096-2485, P.O. Box 2485, Fax +1.201.348-4505, e-mail: orders@springer-ny.com
Eastern Book Service, Japan, Tokyo 113, 3–13, Hongo 3-chome, Bunkyo-ku, Fax +81.3.38 18 08 64, e-mail: orders@svt-ebs.co.jp

Springer-Verlag
and the Environment

WE AT SPRINGER-VERLAG FIRMLY BELIEVE THAT AN
international science publisher has a special obliga-
tion to the environment, and our corporate policies
consistently reflect this conviction.

WE ALSO EXPECT OUR BUSINESS PARTNERS – PRINTERS,
paper mills, packaging manufacturers, etc. – to commit
themselves to using environmentally friendly materials and production processes.

THE PAPER IN THIS BOOK IS MADE FROM NO-CHLORINE
pulp and is acid free, in conformance with international standards for paper permanency.